T0320813

Convex Optimization of Power Systems

Optimization is ubiquitous in power system engineering. Drawing on powerful, modern tools from convex optimization, this rigorous exposition introduces essential techniques for formulating linear, second-order cone, and semidefinite programming approximations to the canonical optimal power flow problem, which lies at the heart of many different power system optimizations.

Convex models in each optimization class are then developed in parallel for a variety of practical applications such as unit commitment, generation and transmission planning, and nodal pricing. Presenting classical approximations and modern convex relaxations side-by-side, and a selection of problems and worked examples, this book is an invaluable resource for students and researchers from industry and academia in power systems, optimization, and control.

Joshua Adam Taylor is Assistant Professor of Electrical and Computer Engineering at the University of Toronto.

Convex Optimization of Power Systems

JOSHUA ADAM TAYLOR

University of Toronto

CAMBRIDGE
UNIVERSITY PRESS

CAMBRIDGE
UNIVERSITY PRESS

University Printing House, Cambridge CB2 8BS, United Kingdom

One Liberty Plaza, 20th Floor, New York, NY 10006, USA

477 Williamstown Road, Port Melbourne, VIC 3207, Australia

314-321, 3rd Floor, Plot 3, Splendor Forum, Jasola District Centre, New Delhi - 110025, India

79 Anson Road, #06-04/06, Singapore 079906

Cambridge University Press is part of the University of Cambridge.

It furthers the University's mission by disseminating knowledge in the pursuit of education, learning and research at the highest international levels of excellence.

www.cambridge.org
Information on this title: www.cambridge.org/9781107076877

First published 2015

A catalogue record for this publication is available from the British Library

Library of Congress Cataloging in Publication data
Taylor, Joshua Adam, 1983–
Convex optimization of power systems / Joshua Adam Taylor.
  pages cm
ISBN 978-1-107-07687-7 (hardback)
1. Electric power systems – Mathematical models.   2. Electric power distribution – Mathematics.   3. Convex programming.   4. Mathematical optimization.   I. Title.
TK1005.T427   2015
621.3101´5196–dc23

                                                                2014032238

ISBN 978-1-107-07687-7 Hardback

For Elodie

Contents

Preface

The application of optimization to power systems has become so common that it deserves treatment as a distinct subject. The abundance of optimization problems in power systems can give the impression of diversity, but in truth most are merely layers on a common core: the steady-state description of power flow in a network. In this book, many of the most prominent examples of optimization in power systems are unified under this perspective.

As suggested by the title, this book focuses exclusively on convex frameworks, which by reputation are phenomenally powerful but often too restrictive for realistic, non-convex power system models. In Chapter 3, the application of classical and recent mathematical techniques yields a rich spectrum of convex power flow approximations ranging from high tractability and low accuracy to slightly reduced tractability and high accuracy. The remaining chapters explore problems in power system operation, planning, and economics, each consisting of details layered on top of the convex power flow approximations. Because all formulations can be solved using standard software packages, only models are presented, which is a departure from most books on power systems. It is a major perk of convex optimization that the user often does not need to program an algorithm to proceed.

I should comment that this book is not an up-to-date exposition of power system applications or optimization theory and that, inevitably, many important topics in both fields have been omitted. My intention has rather been to bridge modern convex optimization and power systems in a rigorous manner. While I have attempted to be mathematically self-contained, the pace assumes an advanced undergraduate level of mathematical exposure (linear algebra, calculus, and some probability) as well as familiarity with power systems and optimization. This book could be used in a course on power system optimization or as a mathematical supplement to a course in power system design, operation, or economics. It is my hope that it will also prove useful to researchers in power systems with an interest in optimization and vice versa, and to industry practitioners seeking firm foundations for their optimization applications.

Acknowledgments

I started this book in the fall of 2011. Most of the present content I learned under the guidance of Franz Hover during my graduate school years at the Massachusetts Institute of Technology and of Duncan Callaway, Kameshwar Poolla, and Pravin Variaya during my postdoctoral studies at the University of California, Berkeley. Certainly, without the open-minded environments they created, this project would have never been attempted. I must also thank a number of colleagues whom I've benefitted from regular discussions with: Eilyan Bitar, Brendan Englot, Reza Iravani, Deepa Kundur, Johanna Mathieu, Daniel Muenz, Ashutosh Nayyar, Andy Packard, Matias Negrete-Pincetic, Anand Subramanian, and many others who have made power systems, control, and optimization such pleasant fields to work in. Finally, Julie Lancashire and Sarah Marsh at Cambridge University Press have made the final stages of this book a highly enjoyable process.

Notation

AC	Alternating current
DC	Direct current
LP	Linear programming
QP	Quadratic programming
SOC(P)	Second-order cone (programming)
SD(P)	Semidefinite (programming)
(C)QCP	(Convex) quadratically constrained programming
MI	Mixed integer
NLP	Nonlinear programming
KKT	Karush-Kuhn-Tucker (conditions)
PNE	Pure strategy Nash equilibrium
MNE	Mixed strategy Nash equilibrium

i	$\sqrt{-1}$
$\mathbb{R}$	The set of real numbers
$\mathbb{C}$	The set of complex numbers
$\mathbb{Z}$	The set of integers
x_i	The ith entry of the vector x
x^k	The kth version of the quantity x, typically corresponding to the kth scenario or time period
Re x	The real part of x
Im x	The imaginary part of x
$|x|$	The absolute value of x
$||x||$	The two-norm of x, $\sqrt{\sum_i x_i^2}$
X_{ij}	The entry at the ith row and j^{th} column of the matrix X
X^T	The transpose of X
X^*	The Hermitian transpose of X. When X is scalar, the complex conjugate.
$X \succeq 0$	The matrix X is positive semidefinite.
rank X	The rank of X
tr X	The trace of X, $\sum_i X_{ii}$
det X	The determinant of X
∇	The gradient operator

To condense exposition, this book employs somewhat relaxed indexing notation. Because there is little risk of ambiguity, i will often be used simultaneously as the imaginary unit and as an index, for example iq_i would be $\sqrt{-1}$ times the ith entry of q. In most cases, constraint indexing will not be explicitly declared; for example,

$$g_i(x) \leq 0$$

is implicitly enforced over $i = 1, \ldots, n$, which is almost always the set of nodes in the network. Similarly, the sum

$$\sum_{ij} x_{ij}$$

is over all relevant node pairs ij, which are usually those connected by lines. Indexing is denoted explicitly when it is not over a standard set, such as when summing over a subset of nodes.

This book makes extensive use of *feasible sets* as organizational tools. Given a collection of constraints $g_i(x) \leq 0$, the corresponding feasible set is

$$\{x \mid g_i(x) \leq 0\},$$

i.e., the set of points for which every constraint is satisfied.

1 Introduction

1.1 Recent history

Streetlights, subways, the Internet, this book you are reading now – it is difficult to imagine life without such amenities, all enabled by electric power. To support such a vast set of technologies, electric power systems have grown into some of the most complex and expensive machines in existence. While much of this growth resembles an organic process more than deliberate design, the advent of computing is enabling us more and more to direct the evolution of power systems toward greater efficiency, reliability, and versatility.

At the time of writing, the complexity of power systems is poised to take off. This is largely due to shifts toward renewable energy production and the active involvement of power consumers through demand response, as well as our still-developing handle on economic deregulation. To meet these challenges, new computational tools will be developed, and the most ubiquitous computation in power systems is optimization. An objective of this book is to simplify and unify various topics in power system optimization so as to provide a firm foundation for future developments.

At the heart of most power system optimizations are the equations of the steady-state, single-phase approximation to alternating current power flow in a network. Well-known problems like optimal power flow, reconfiguration, and transmission planning all consist of details layered on top of power flow. Nodal prices, a core component of electricity markets, are obtained from the dual of optimal power flow. It is therefore most unfortunate that the power flow equations are nonconvex, making all of these optimizations extremely difficult. We are thus faced with a tradeoff between realistic models that are too hard to solve at practical scales and tractable approximations.

For many years, linear programming (LP) was the most general efficiently solvable optimization class, and so many large-scale power system models were based on linear power flow approximations or even simpler descriptions like network flow or a real power balance. At the other extreme, a number of nonlinear programming (NLP) algorithms were developed for exact, nonconvex models. These approaches invariably encountered difficulty scaling to larger problem sizes due to the underlying NP-hardness of nonconvex optimization. This led some to resort to so-called metaheuristic algorithms, which make little use of problem structure and give little indication of their performance. Beyond their scalability issues, NLP and metaheuristic approaches can be tiresome to implement because they often require the user to program both the mathematical model and algorithm. On the other hand, one would rarely write their own algorithm to solve an LP because of the many available professional-grade commercial and academic implementations.

In 1984, there was a turning point for convex optimization when Karmarkar invented the first practical, polynomial-time interior point method for LP [1]. Over the next decade, second-order cone programming (SOCP) and semidefinite programming (SDP) emerged as convex generalizations of LP also admitting polynomial-time interior point methods [2]. Of equal importance, SOCP and SDP are now featured in a number of standard software packages, making them similarly user friendly.

The enhanced modeling capabilities of SOCP and SDP brought about an explosion of research applications, some of which can be found in the standard text [3]. In 2006, ripples from the previous twenty years were felt in power systems when power flow in radial networks was posed as an SOCP in Jabr [4] and again in 2008 when an SDP approximation of optimal power flow was developed in Bai, Wei, Fujisawa, and Wang [5]. A substantial body of research has materialized in the short time since then, both theoretically characterizing the new SOCP and SDP power flow approximations and applying them in a variety of power system contexts. For most power system optimization problems, we now have a spectrum of LP, SOCP, SDP, and NLP models to choose from, each with a different balance of realism and scalability.

1.2   Structure and outline

This book only contains models and, with the exception of Chapters 2 and 7, rarely mentions algorithms. This is *not* because the algorithms are not worth knowing or decoupled from modeling; one can *always* do better by formulating optimization models and algorithms jointly. Rather, here this wisdom is applied by formulating models so that they can be solved by certain algorithms. This approach is a luxury we can afford because optimization is a relatively mature field: for a desired level of scalability, it identifies the corresponding tradeoff between efficiency and descriptiveness. Here, this manifests as a hierarchy of convex optimization classes, the main elements of which are LP, SOCP, and SDP. LP is the most efficient and least descriptive, SDP vice versa, and SOCP is in between. As discussed in the previous section, once a model has been formulated within one of these classes, it can be conveniently solved using standard software packages. By tailoring our models to these classes, we arrive at tractable formulations far more easily than if we were to design both model and algorithm from scratch.

The resulting separation between models and algorithms should not be seen as a restriction but as a starting point for further specialization and extension. For example, once a problem has been formulated as an LP, uncertainty can be mechanistically incorporated using robust optimization (Section 7.1.2), or the problem can be split into different chunks for multiple processors or agents (Section 7.2).

A central motif in this book is posing complicated models as layers on top of more basic ones. To avoid rewriting the same constraints over and over, this book makes extensive use of *feasible sets* to package frequently occurring groups of constraints for concise representation in other problems. Consequently, certain parts of this book are highly cumulative. Each chapter is concluded by a summary highlighting important points and open problems, as well as a small selection of exercises. In addition to those given, which are all analytical, students are of course encouraged to implement each chapter's models and examples using their preferred optimization platform.

The chapter structure is summarized below.

Chapter 2: This chapter provides a minimal introduction to optimization and power system modeling. In particular, it defines LP, SOCP, and SDP, and the tools used to construct convex relaxations. It also derives the quadratic steady-state, single-phase approximation to power flow in a network.

Chapter 3: This chapter defines the basic, nonconvex optimal power flow problem. It then derives classical linear approximations and modern SOC and SD relaxations. The constraints in these models form the foundation of all subsequent chapters in this book.

Chapter 4: This chapter constructs linear, SOC, and SD versions of a number of central optimization problems in power system operations using the approximations from the previous chapter. Here and in the next chapter we encounter a number of mixed-integer constraints, which make these problems challenging even with linear power flow constraints. This chapter also takes brief detours though inventory control and linear quadratic regulation.

Chapter 5: Similar to the last chapter, this chapter constructs infrastructure planning problems around the power flow approximations of Chapter 3. In this chapter, every problem is extremely difficult due to the integer constraints required to describe component installation.

Chapter 6: This chapter discusses electricity markets, in which *nodal prices* are obtained from the dual of optimal power flow. Because each convex approximation has strong duality, prices are guaranteed to support *economic dispatch*. This chapter also discusses why basic economic assumptions never hold in practice and briefly summarizes some game theoretic analyses of market power.

Chapter 7: This final chapter surveys some promising directions for future work.

1.3   On approximations

Every model in this book is an approximation. In fact, every mathematical model ever is an approximation, which inevitably fails to capture some fine physical detail. However, it is worth stating this explicitly here because every model in this book involves an approximation of one particular model: the steady-state, single-phase description of electric power in a network of conductors. We derive this model in Section 2.5 and place it in the context of optimization in Section 3.1.

The steady-state, single-phase description of electric power is a special approximation because it derives from very natural physical assumptions and enables power system engineers to take a large step from simulation to design, which is what this book is all about. By definition, optimization is the highest (mathematical) form of design. Unfortunately, the steady-state description of electric power isn't quite right for optimization because it is nonconvex. This book attempts to make the mildest further adjustments necessary for this model and those built atop it to enjoy all that optimization has to offer.

With this perspective in mind, it is helpful to remember the following two statements when using this book.

- We would always use a more realistic description of electric power like an unbalanced steady-state or transient model were it practical to do so. So, when it is practical, use one of them and not the convex approximations in this book. We briefly elaborate on this in Section 3.1.1.
- While immensely important, the steady-state description of electric power is not sacrosanct. It is an approximation like all of the other models in this book, which just happens to be (slightly) closer to reality. From this perspective, the convex approximations in this book are no less valid than the model from which they are derived.

References

[1] N. Karmarkar, "A new polynomial-time algorithm for linear programming," in *Proceedings of the Sixteenth Annual ACM Symposium on Theory of Computing*, ser. STOC '84. New York: ACM, 1984, pp. 302–311.

[2] Y. Nesterov and A. Nemirovski, "Interior point polynomial methods in convex programming," *SIAM Studies in Applied Mathematics*, vol. 13, 1994.

[3] S. Boyd and L. Vandenberghe, *Convex Optimization*. New York: Cambridge University Press, 2004.

[4] R. Jabr, "Radial distribution load flow using conic programming," *IEEE Transactions on Power Systems*, vol. 21, no. 3, pp. 1458 –1459, Aug. 2006.

[5] X. Bai, H. Wei, K. Fujisawa, and Y. Wang, "Semidefinite programming for optimal power flow problems," *International Journal of Electrical Power and Energy Systems*, vol. 30, no. 6–7, pp. 383–392, 2008.

2   Background

This chapter summarizes the basic technical concepts used throughout this book. As stated in the Introduction, this book focuses on modeling, so most algorithmic aspects are left "under the hood." Because this book is intended to appeal to anyone familiar with power systems or optimization, background material on both topics is covered, albeit at the minimum depth necessary to access the later material.

2.1   Convexity and computational complexity

We begin with a few core concepts. A point $x_0 \in X$ is a *global minimum* of the function $f(x)$ over the set $X \subseteq \mathbb{R}^n$ if $f(x_0) \leq f(x)$ for all $x \in X$. If $f(x)$ is continuous and X is compact, which is to say closed and bounded, such a point is guaranteed to exist. x_0 is a *local minimum* of $f(x)$ if there exists an $\epsilon > 0$ for which $f(x_0) \leq f(x)$ for all $x \in X$ satisfying $\|x - x_0\| \leq \epsilon$. All minima of a convex function achieve the same function value and are therefore global. In general, a function may have multiple local and global minima.

To find a global minimum of a convex function, choose a descending algorithm, let it run free, and in a perfect world it will eventually end up there. (In the real world, large problem sizes or bad numerical conditioning can derail any algorithm.) The intuitive simplicity of convexity translates to a genuine computational advantage, which is evinced by the powerful algorithms that exist for convex optimization and the extreme difficulty of nonconvex optimization. This section gives some basic characterizations of convexity and describes the varieties of convex optimization problems encountered in this book. For more comprehensive coverage, the reader is referred to endnotes [1–5], and to endnotes [6, 7] for theoretical treatments of convex functions and sets.

A function f is convex if, for any two points in its domain, x and y,

$$f(\alpha x + (1 - \alpha)y) \geq \alpha f(x) + (1 - \alpha)f(y) \quad \text{for all } \alpha \in [0, 1].$$

This means that any point on the straight line between $(x, f(x))$ and $(y, f(y))$ is greater than or equal to the function value at the corresponding point between x and y. When f is twice-differentiable, it is convex if and only if its Hessian is positive semidefinite:

$$\nabla^2 f(x) \succeq 0.$$

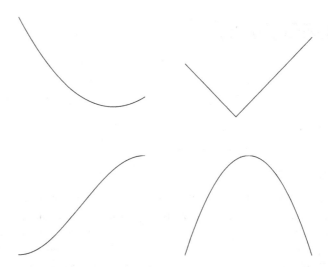

Fig. 2.1   Convex (top) and nonconvex (bottom) functions. The top left function is strictly convex, and the top right is not. The bottom right function is strictly concave.

f is strictly convex if the $\geq$ in the first condition and $\succeq$ in the second are respectively replaced by $>$ and $\succ$. A strictly convex function has a unique global minimum. Some single variable convex and nonconvex functions are shown in Figure 2.1.

A set X is convex if, for any x and y in X, any point $\alpha x + (1 - \alpha)y$ with $\alpha \in [0, 1]$ is also in X; in other words, all points on the line between x and y are also in X. Convex and nonconvex sets are shown in Figure 2.2. Now consider the set $X = \{x \mid g(x) \leq 0\}$; if g is convex, then so is X. While extremely useful in general, we will virtually never use these characterizations of convexity in this book because we will seek models that are convex by construction and hence do not need to be tested.

Suppose that we want to solve the following optimization problem:

$$\underset{x}{\text{minimize}}\ f(x)$$
$$\text{subject to}\ \ g_i(x) \leq 0.$$

If $f(x)$ and each $g_i(x)$ are convex, any minimal solution is a global minimum attaining the lowest possible function value, and we call the above a *convex optimization problem*. We say that a point x is *feasible* if $g_i(x) \leq 0$ for all i, i.e., all constraints are satisfied. Equality constraints $h(x) = 0$ are equivalently expressed by the pair $h(x) \leq 0$ and $h(x) \geq 0$, from which it is evident that the only convex equality constraints are linear.[1]

In power system optimization problems, the objective function is usually some economic cost, and the constraints encode physical requirements derived from first principles. Consequently, we will usually assume approximate, convex objectives while regarding nonconvex constraints as fundamental. A primary focus of this book is therefore the nonconvex constraints arising in power system optimization problems, which

---

[1] Technically, a linear function is of the form $g(x) = \alpha x$. In optimization, it is standard to use the term "linear" to refer to affine functions of the form $g(x) = \alpha x + \beta$.

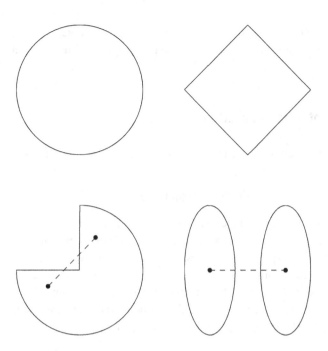

Fig. 2.2   Convex (top) and nonconvex (bottom) sets. Convexity violations are shown for the latter.

are organized into *feasible sets*. The feasible set of an optimization problem is the set of all feasible points, i.e.,

$$\{x \mid g_i(x) \le 0\}.$$

Convex optimization problems are often efficiently solvable because it is unnecessary to check numerous local minima to find a global minimum. This sort of computational efficiency can be quantified by the concept of non-deterministic polynomial-time hardness, or NP-hardness for short [8–10] (actually, some of the problems we will cover are NP-complete, a subset of the NP-hard problems, but the difference is insignificant for our purposes). On a fundamental level, there is no way to efficiently solve a large instance of an NP-hard problem.

More precisely, suppose the difficulty of a problem grows with some parameter, n. A *polynomial-time* algorithm will solve said problem (terminate with a satisfactory answer) after a number of arithmetic operations, which is a polynomial in n, for example n^q, where q is some positive integer. An *exponential-time* algorithm will require a number of operations that is exponential in n; if you are unconvinced that this is a significant difference, compare q^n and n^q for any fixed q and increasing n (alternatively, try counting to 2^{100}).

In the context of optimization, a polynomial-time algorithm requires a number of operations that is polynomial in both the number of variables and constraints. One reason nonconvex problems can be NP-hard is that the number of local minima may be exponential in either the number of constraints or variables. Although convexity is not the precise boundary between hard and easy problems, there are broad classes of convex optimization problems that admit polynomial-time algorithms, and nonconvex ones usually do not.

2.2   Optimization classes

This section describes the classes of optimization problems encountered in this book.

2.2.1   Linear and quadratic programming

LP rightfully occupies the central spot among efficient convex optimization frameworks; the reader is referred to endnotes [11–14] for comprehensive treatments. Even Dantzig's simplex algorithm, which is known to have worst-case exponential complexity, remains the most competitive choice for many LPs and is at the foundation of mixed-integer algorithms. QP is LP with an additional quadratic term in the objective, and when the quadratic term is convex, QPs can be solved with similar algorithms and efficiency to LPs. The *standard form* of a generic QP is

$$\underset{x}{\text{minimize}} \; x^T Q x + c^T x$$
$$\text{subject to} \; Ax = b$$
$$x \geq 0.$$

This is the format processed by most algorithms. Observe that if the rank of A is equal to the length of the vector x, $Ax = b$ is a fully determined system with only one solution. If it is positive, it is feasible and therefore optimal. If $Ax = b$ is overdetermined, no feasible solution exists and the problem is also trivial. It is only when rank A is less than the dimension of x that the feasible set is more than a point and the above optimization is nontrivial. Because all linear constraints are convex, the feasible set of an LP or QP is always convex. The feasible set of a simple LP is shown in Figure 2.3.

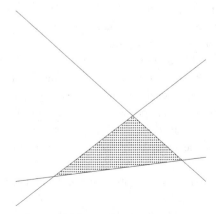

Fig. 2.3   The (convex) polyhedral feasible set described by an LP's inequality constraints.

When Q is a positive semidefinite matrix (which we define in Section 2.2.2 and denote $Q \succeq 0$), this problem is convex. When Q is indefinite or negative semidefinite, this problem is NP-hard [15]. When $Q = 0$, we obtain the standard form LP. All three cases are illustrated in Example 2.1. Putting a model into standard form can be tedious, but fortunately many software-modeling environments exist to automate this task.

The constraints of an LP or QP constitute a convex *polytope*, a set with only flat sides. The simplex algorithm operates by descending the corners or *vertices* of the polytope. In 1984, Karmarkar introduced the first practical polynomial-time LP algorithm [16]. It was an interior point method that, unlike the Simplex method, used Newton's method to navigate the interior of the feasible set and avoided constraints via barrier functions. This was a milestone in optimization and paved the way for polynomial-time algorithms for more general classes of convex optimization problems, in particular SOCP and SDP.

Example 2.1 *A small quadratic program.* Consider a two-dimensional QP with standard form parameters

$$Q = \begin{bmatrix} q & 0 \\ 0 & 0 \end{bmatrix}, \quad c = \begin{bmatrix} 1 \\ 0 \end{bmatrix}, \quad A = [1 \quad 1], \quad b = 1.$$

Observe that these parameters encode the one dimensional, non-standard form QP

$$\underset{y}{\text{minimize}} \quad qy^2 + y$$

$$\text{subject to } 0 \le y \le 1.$$

If $q \ge 0$, $Q \succeq 0$, and the local minimum $x = [0; 1]$ is the global minimum due to convexity. If $q = 0$, the problem is linear, and the solution is again $x = [0; 1]$. Now suppose that $q = -2$. Then both $x = [0; 1]$ and $x = [1; 0]$ are local minima, and we can ascertain that the latter is the global minimum by comparing the corresponding objective values.

Observe that the only way to find the global minimum when $q = -2$ was to check both local minima. This kind of enumerative procedure may be straightforward in low-dimensional problems but would hardly be viable if x was a thousand-dimensional vector hiding in a polytope with more than 2^{1000} vertices to check. In fact, no efficient procedure exists for general problems of this nature because they are NP-hard.

Now observe that if $-1/2 \le q \le 0$, the problem is nonconvex, but $x = [0; 1]$ is the sole local and hence global solution. We are here reminded that convexity is a sufficient condition for global optimality of local optima but not a necessary one.

2.2.2 Cone programming

LP is the most basic type of cone programming. The constraint $x \ge 0$ says that x must occupy the cone of points with all nonnegative coordinates, also known as the nonnegative orthant. More generally, a set X is a cone if for any $x \in X$ and nonnegative scalar α, $\alpha x \in X$. It is a convex cone if it also satisfies the definition of set convexity in Section 2.1.

This section discusses two convex conic optimization classes: *second-order cone* and *semidefinite programming*. They are successive generalizations of LP and share with it the virtue of polynomial-time interior point methods (but not practical simplex methods). Of course, with greater generality comes lower efficiency, so one would never solve an LP as an SOCP or an SOCP as an SDP.

Second-order cone programming

The SOC, also referred to as the Lorenz or ice cream cone, is defined as the set

$$\left\{(y,t) \in \mathbb{R}^{n+1} : \|y\| \leq t\right\}.$$

Clearly, if $\|y\| \leq t$, then $\|\alpha y\| \leq \alpha t$ for any $\alpha \geq 0$, so it satisfies the definition of a cone. Detailed expositions of SOCP are provided in endnotes [17, 18]. We can straightforwardly check its convexity. Suppose (y_1, t_1) and (y_2, t_2) are in the SOC and $\alpha \in [0, 1]$. Then, using the triangle inequality and homogeneity of the two-norm,

$$\|\alpha y_1 + (1 - \alpha)y_2\| \leq \alpha \|y_1\| + (1 - \alpha) \|y_2\|$$
$$\leq \alpha t_1 + (1 - \alpha)t_2,$$

i.e., the point $(\alpha y_1 + (1 - \alpha)y_2, \alpha t_1 + (1 - \alpha)t_2)$ is also in the SOC.

Setting $y = Ax + b$, $A \in \mathbb{R}^{m \times n}$, $b \in \mathbb{R}^m$, and $t = c^T x + d$, $c \in \mathbb{R}^n$, $d \in \mathbb{R}$, we obtain the standard form SOC constraint:

$$\|Ax + b\| \leq c^T x + d. \tag{2.1}$$

An example SOC constraint is shown in Figure 2.4. Notice that if A is zero, this is a linear constraint. If $c = 0$, CQCP is retrieved, which we discuss further in Section 2.2.3.

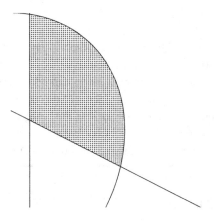

Fig. 2.4   The feasible set described by the linear constraints $x_1 \geq 0$, $1 - x_1 \leq 2x_2$, and an SOC constraint with the standard form parameterization $A = [1, 2; -1, 1]$, $b = [1, -1]^T$, $c = [1, 1]^T$, $d = 2$. Here, x_1 and x_2 are the horizontal and vertical axes, respectively.

The standard form SOCP is

$$\underset{x}{\text{minimize}} \ k^T x$$

$$\text{subject to} \ \|A_i x + b_i\| \le c_i^T x + d_i.$$

Semidefinite programming

The most general class of convex cone programs appearing in this book is SDP [19]. Positive semidefiniteness is a generalization of scalar nonnegativity to Hermitian matrices.[2] There are many equivalent characterizations of positive semidefiniteness, including:

- $z^* X z \ge 0$ for all $z \in \mathbb{C}^n$.
- All eigenvalues of X are nonnegative.
- All principal minors of X are nonnegative.[3]

If $X \succeq 0$ and $\alpha \ge 0$, then $\alpha X \succeq 0$, so the set of positive semidefinite matrices is a cone. In-depth treatments of positive semidefiniteness and related topics in linear algebra are provided in Horn and Johnson [20]. Again, convexity is easy to show. Suppose X_1 and X_2 are in the SD cone, i.e., $X_1 \succeq 0$ and $X_2 \succeq 0$, and $\alpha \in [0, 1]$. Then, for any $z \in \mathbb{C}^n$,

$$z^* \left(\alpha X_1 + (1 - \alpha) X_2 \right) z = \alpha z^* X_1 z + (1 - \alpha) z^* X_2 z$$
$$\ge 0,$$

which establishes that $\alpha X_1 + (1 - \alpha) X_2 \succeq 0$. A useful consequence of the definition of positive semidefiniteness is that all principal submatrices of a positive semidefinite matrix are also positive semidefinite. Analogous characterizations of positive definiteness, the "strict" version of positive semidefiniteness, are obtained by replacing $\ge$ with $>$ and the word "nonnegative" with "positive."

We denote the condition "X is positive semidefinite" by the notation $X \succeq 0$. We will sometimes indicate "$X - Y$ is positive semidefinite" with the notation $X \succeq Y$. Optimization variables often appear in semidefinite constraints as linear matrix functions of the form

$$F(x) = F_0 + F_1 x_1 + \cdots + F_n x_n \succeq 0.$$

For this reason, semidefinite constraints are commonly referred to as *linear matrix inequalities* [21].

Example 2.2 *Eigenvalue optimization.* A number of important engineering problems can be cast as eigenvalue optimizations, in which some function of a matrix's

---

[2] A matrix X is Hermitian if it is equal to its conjugate transpose, $X = X^*$.
[3] A minor is the determinant of a submatrix of X. It is a principal minor if the submatrix's main diagonal lies on the main diagonal of the full matrix.

eigenvalues is minimized. Examples include truss topology design [22], Markov chain mixing times [23], and consensus algorithms [24].

Suppose the Hermitian matrix $A(x)$ depends affinely on x and that we want to maximize its minimum eigenvalue, λ, over x. λ can be defined in terms of the Rayleigh quotient as

$$\lambda = \min_{y} \frac{y^* A(x) y}{y^* y},$$

which is equivalently stated as the largest λ such that

$$y^* (A(x) - \lambda I) y \geq 0 \quad \text{for all } y.$$

But this is exactly one of the given definitions of positive semidefiniteness, now applied to the matrix $A(x) - \lambda I$. We can thus encode the condition "λ is no larger than the minimum eigenvalue of $A(x)$" as an SD constraint, which gives us the SDP

$$\begin{aligned} \underset{x,\lambda}{\text{maximize}} \quad & \lambda \\ \text{subject to} \quad & A(x) \succeq \lambda I. \end{aligned}$$

Likewise, the maximum eigenvalue is minimized via

$$\begin{aligned} \underset{x,\lambda}{\text{minimize}} \quad & \lambda \\ \text{subject to} \quad & A(x) \preceq \lambda I. \end{aligned}$$

Note that, in general, maximizing the maximum or minimizing the minimum eigenvalue cannot be solved by SDP. Also note that the sum of the eigenvalues is simply $\operatorname{tr} A(x)$, which is linear and thus can be optimized with LP.

A useful construction in SDP is the *Schur complement*. Given three matrices, $A \in \mathbb{R}^{n \times n}$, $B \in \mathbb{R}^{n \times m}$, and $C \in \mathbb{R}^{m \times m}$, the Schur complement of the matrix

$$D = \begin{bmatrix} A & B \\ B^* & C \end{bmatrix}$$

is defined as

$$S = C - B^* A^{-1} B.$$

Now suppose that A is positive definite. Then $S \succeq 0$ if and only if $D \succeq 0$. Similarly, if C is positive definite, then $A - B C^{-1} B^* \succeq 0$ if and only if $D \succeq 0$. We make immediate use of this property by showing that SOCP is indeed a special case of SDP.

Example 2.3   *SOCP $\subset$ SDP.* Arrange the real vector $Ax + b$ and scalar $c^T x + d$ in one positive semidefinite matrix as

$$\begin{bmatrix} I & Ax + b \\ (Ax + b)^T & (c^T x + d)^2 \end{bmatrix} \succeq 0.$$

Because the identity, I, is positive definite, the Schur complement must also be positive semidefinite, so that

$$(Ax + b)^T (Ax + b) \leq \left(c^T x + d\right)^2. \tag{2.2}$$

Taking the positive square root of both sides, we recover the standard form SOC constraint, (2.1). This establishes that any SOC constraint is representable as an SD constraint and therefore that SDP is a generalization of SOCP. Of course, SOCP is a more tractable optimization class than SDP, and therefore SOCPs should never be solved as SDPs.

As with LP and SOCP, there is a standard form for SDP, which is given by

$$\underset{X}{\text{minimize}} \quad \text{tr} \; C^* X$$

$$\text{subject to} \quad \text{tr} \; A_i^* X \leq b_i$$

$$X \succeq 0.$$

Note that while perhaps unconventional in appearance due to the trace operator, the objective and first set of constraints are linear; it is only $X \succeq 0$ that makes this an SDP. We will, however, not employ this format in this book because it is not the most convenient for our modeling purposes, and, as with LP, existing software packages convert SDPs to standard form automatically.

This section has presented SDP in terms of complex matrices because one of the central optimization problems in this book is most conveniently formulated as a complex SDP. Despite the fact that most optimization routines work exclusively in real numbers, this is not a limitation because any complex SDP can be equivalently represented as a real SDP with slightly more variables.

2.2.3 Quadratically constrained programming

QCP is the broadest class of optimization problems encountered in this book. In fact, the bulk of this book is concerned with constructing LP, SOCP, and SDP approximations of QCPs arising in power systems. This approach is useful because QCP is NP-hard and hence very difficult.

Because QCP is such a large and thorny optimization class, there are no specialized algorithms for solving general QCPs as there are for LP, SOCP, and SDP. Consequently, there is little value in having a standard form for QCP. A generic QCP can be written

$$\underset{x}{\text{minimize}} \quad x^* Q x + c^* x$$

$$\text{subject to} \quad x^* R_i x + r_i^* x + d_i \leq 0.$$

First, note that we allow the variables and parameters to be complex. As with SDP, this is purely a notational issue; so long as the objective and constraints are real-valued, any complex QCP can be written as a real QCP by defining additional variables. Because voltage in steady-state is usually represented by a complex phasor (see Section 2.5), it will be convenient for us in Chapter 3 to formulate our basic models as complex QCPs.

If $Q \succeq 0$ and each $R_i \succeq 0$, the QCP is convex and is a special case of SOCP, as was mentioned in Section 2.2.2. Because SOCP is a relatively slight generalization of CQCP, SOCP algorithms are generally used to solve CQCPs. Example 2.4 shows that SOCP is in fact contained by the more general class of QCPs.

Example 2.4 *SOCP $\subset$ QCP.* Recall the standard form SOC constraint:

$$\|Ax + b\| \leq c^T x + d.$$

By simply squaring and expanding both sides, this can be expressed as the quadratic inequality

$$x^T \left(A^T A - c c^T \right) x + 2 \left(b^T A - d c^T \right) x + b^T b - d^2 \leq 0,$$

which is a QCP constraint. We can now clearly see that, as mentioned in Section 2.2.2, CQCP is equivalent to SOCP with $c = 0$ because $A^T A$ must be positive semidefinite when A is real.

Note that even if the matrix $A^T A - c c^T$ is not positive semidefinite, the constraint is guaranteed to be convex by virtue of being an SOC constraint. This example highlights an interesting phenomenon wherein a problem can be nonconvex when written one way and convex another. We will find over the course of this book that many such nonconvex QCPs are very nearly convex when written the right way, enabling us to construct high-quality linear, SOC, and SD approximations.

Nonconvexity imparts a remarkable degree of generality to QCP. Because CQCP fits relatively snugly inside of SOCP, we know somewhat intuitively the limits of its descriptive capabilities. It is then surprising how descriptive QCP is; for instance, the next section shows that the notoriously difficult class of integer programs is merely a special (nonconvex) case of QCP. We conclude our discussion of QCP by showing that SDP is a special case of QCP.

Example 2.5 *SDP $\subset$ QCP.* To show that SDP is contained in QCP, we can show that any standard form SDP can be written as a QCP. This is equivalent to expressing $X \succeq 0$ as a collection of quadratic inequality constraints because the standard form SDP objective and other constraints are linear and hence within QCP.

Recall from Section 2.2.2 that $X \succeq 0$ if and only if all of its principal minors are nonnegative. $X \succeq 0$ is then equivalent to enforcing the nonnegativity of a finite number of principal minors. But a determinant is a polynomial function of the entries of X, and any polynomial constraint can be written as a quadratic constraint by introducing more variables and constraints. (For example, $x_1 x_2 x_3 \leq 0$ is equivalent to the pair of inequalities $x_1y \leq 0$ and $x_2x_3 \leq y$.) Therefore, the SD constraint $X \succeq 0$ may be equivalently written as a finite number of quadratic inequalities.

By showing that the standard form of SDP can be written as a (likely nonconvex) QCP, we have shown that the class of SDPs is indeed contained in QCP. As in Example 2.4, this reinforces the wisdom that the convexity of a problem can depend on how it is written. However, just as an LP shouldn't be solved as an SOCP, an SDP should be solved as an SDP and not as a QCP.

2.2.4   Mixed-integer programming

Integer constraints arise naturally when modeling things that only make sense in discrete quantities. For example, a switch can only be open or closed, a transmission line from New York to Boston cannot be a micron thick, and usually a power plant is either there or it isn't. Integer constraints pervade electric power systems. Refer to endnotes [9, 13, 25–27] for in-depth coverage of integer and mixed-integer programming.

We express integer constraints simply by stating that a variable is in the set of integers, or, if it can only take on two values (like "open" and "closed"), it must be zero or one. We respectively write these conditions

$$x \in \mathbb{Z} \quad \text{and} \quad x \in \{0, 1\}.$$

Figure 2.5 shows a feasible set in which all variables are constrained to the integers.

In this book, we will primarily encounter *mixed-integer programs* in which some variables are discrete and others continuous. We will seek mixed-integer programming formulations that have convex *continuous relaxations*, which are convex if all integer constraints are removed. Consequently, the same approximations will be used regardless of the presence of integer constraints. Nothing further will be done to approximate integer constraints because powerful heuristics exist for mixed-integer programs with convex continuous relaxations. Mixed-integer programming is illustrated in Example 2.6 via the disjunctive constraint, a modeling tool we will use frequently.

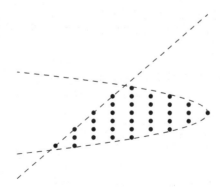

Fig. 2.5   A discrete feasible set described by integrality and continuous inequality constraints.

Example 2.6 *Disjunctive constraints.* Consider the following optimization problem:

$$\underset{x}{\text{minimize}} \; f(x)$$

$$\text{subject to} \; g_1(x) \leq 0 \quad \text{or} \quad g_2(x) \leq 0.$$

If the "or" condition was replaced by "and," both constraints could be enforced, but instead we must accept solutions that satisfy at least one but not necessarily both. Optimization with logical constraints of this sort is sometimes referred to as disjunctive programming [28, 29].

This problem may be written as an ordinary optimization problem by introducing a new binary variable ξ and rewriting it as

$$\underset{x,\xi}{\text{minimize}} \; f(x)$$

$$\text{subject to} \; g_1(x) \leq M\xi$$

$$g_2(x) \leq M(1 - \xi)$$

$$\xi \in \{0, 1\}.$$

The parameter M must be large enough that, if $\xi = 1$, the first constraint does not influence the solution, and likewise when $\xi = 0$ for the second constraint. A simple disjunctive constraint is shown in Figure 2.6.

This approach can accommodate a diverse range of scenarios in which multiple constraints are dependent through a logical condition, and will be used frequently over the course of this book.

While convex conic optimization situates a broad spectrum of important problems within the scope of standardized polynomial-time algorithms, integer constraints will always be an inescapable source of NP-hardness. Intuitively, one might say that there are fewer integers than real numbers, and therefore they should be easier to optimize over, but integer constraints must be challenging for the following reason: one can explore a

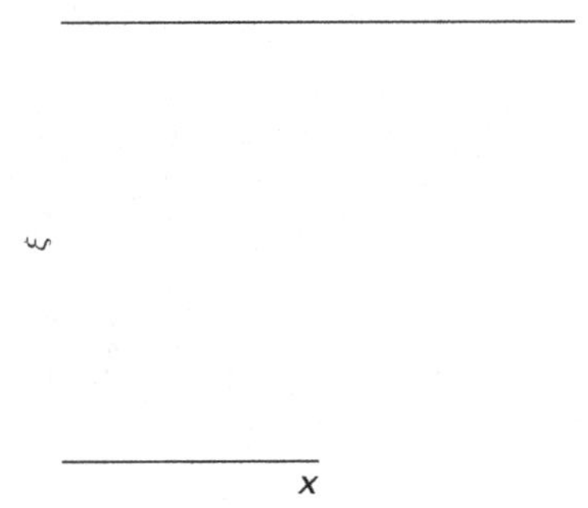

Fig. 2.6 The feasible set associated with the mixed-integer disjunctive constraint $0 \leq x \leq 1 - \xi/2$, $\xi \in \{0, 1\}$ is composed of two horizontal lines of different lengths.

continuous function without checking every point via gradients and Hessians, but the only way to ascertain the quality of an integer solution is to go there. In other words, continuous functions can be subjected to directed searches, but discrete points must ultimately be enumerated.

Because integer programming is NP-hard and hence will always require an exponential amount of time in the worst case, we can only use heuristics to try to accelerate convergence. The two most popular are branch-and-bound and cutting planes. Branch-and-bound is a general tool applicable to all integer programming problems, while practical cutting planes presently only exist for MILP and MIQP. However, MISOCP is currently a highly active area of research, and practical, standardized algorithms are to be expected in the near future, cf. [30–32].

We have devoted an entire section to integer constraints because they are ubiquitous and necessitate specialized modeling and algorithmic treatments. However, we implicitly covered integer constraints in Section 2.2.3 in our discussion of nonconvex quadratic constraints. Consider the binary constraint $x \in \{-1, 1\}$. This is equivalently written $x^2 = 1$, which is identical to the pair of quadratic inequality constraints

$$x^2 \leq 1 \quad \text{and} \quad x^2 \geq 1.$$

Observe that only the second constraint is nonconvex. Now consider a generic integer variable, $z \in \mathbb{Z}$, $0 \leq z \leq \bar{z}$. This is equivalently expressed

$$z = \sum_{k=0}^{\bar{z}} k y_k, \quad \sum_{k=0}^{\bar{z}} y_k = 1, \quad y_k \in \{0, 1\}.$$

Hence, because any integer constraint can be written as a collection of binary and linear constraints, all of integer programming is contained by QCP.

2.2.5 Algorithmic maturity

We now have presented successively more general convex and nonconvex optimization frameworks, the breadths of which are summarized by the following (proper) inclusions:

$$\underbrace{\text{LP} \subset \text{QP} \subset \text{CQCP} \subset \text{SOCP} \subset \text{SDP}}_{\text{Convex, solvable in polynomial-time}} \subset \underbrace{\text{QCP} \subset \text{NLP}}_{\text{Nonconvex, NP-hard}}$$

The classes on the left are the least descriptive and most efficiently solvable, and vice versa for those on the right. We must thus straddle the trade-off between modeling tractably on the left and realistically on the right. This book implements this by approximating "exact" models in the rightmost classes with tractable, convex conic models in the others. Table 2.1 summarizes the maturity and types of algorithms available for each conic optimization framework.

The simplex algorithm can solve LPs and QPs with convex objectives extraordinarily fast. They can be solved with similar efficacy using interior point methods. Specialized interior point methods can also solve SOCPs and SDPs with high theoretical

Table 2.1 Present-day algorithms and maturity of each convex conic optimization class. (IP: interior point method, B&B: branch and bound, CP: cutting planes)

Problem class	Algorithms		Maturity	
	Continuous	Integer	Continuous	Integer
LP/QP	Simplex, IP	B&B, CP	High	High
CQCP/SOCP	IP	B&B, CP	High	Moderate
SDP	IP	B&B	Moderate	Low

(polynomial-time) efficiency, but, at the time of writing, this only translates to practical efficiency for SOCP. However, recent history suggests that it will not be long before SDP algorithms attain efficiency and robustness commensurate with SOCP.

Informally, integer constraints make any problem exponentially more difficult. Despite being NP-hard, MILPs and MIQPs can nevertheless be solved quite effectively using heuristic algorithms discussed in the previous section. MISOCP algorithms are very much in progress, and MISDP may be further off.

While the present-day utility of each optimization class is determined by its existing software implementations, it is their polynomial-time guarantees or lack thereof that shape our future expectations. Efficient approaches may never exist for the general class of QCPs, but it is very reasonable to expect powerful algorithmic implementations for all of the above continuous convex frameworks and decent heuristics for their mixed-integer counterparts. It is with this perspective that the models are formulated in this book.

2.3   Relaxations

Almost all basic formulations of power system optimization problems are nonconvex. Traditionally, this has been circumvented through linearization, which is covered in Section 3.2. While linear approximations have been and will continue to be indispensable to power system design and operation, tractable but more accurate options now exist. Our primary vehicle is the relaxation.

Consider the optimization:

$$\text{minimize } f(x) \quad \text{subject to} \quad x \in X.$$

The optimization

$$\text{minimize } f(x) \quad \text{subject to} \quad x \in Y$$

is a relaxation if $X \subseteq Y$ (i.e., for any $x \in X$, $x \in Y$).

The above definition is illustrated in Figure 2.7. Clearly, a relaxation is only meaningful if there are $x \in Y$, $x \notin X$, since otherwise the two are identical. A consequence of this definition is that the minimum objective of a relaxation is less than or equal to

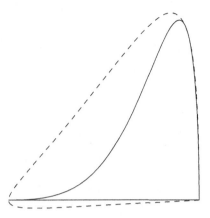

Fig. 2.7   A convex relaxation (dashed) of a nonconvex set (solid).

that of the original problem.[4] One of our basic approaches is to approximate nonconvex problems with convex relaxations.

Relaxations are superior to general approximations because they provide bounds on true optima. Suppose $X \subset Y$. Then, by construction,

$$\min_{x \in Y} f(x) \leq \min_{x \in X} f(x) \leq f\left(x'\right), \; x' \in X.$$

A relaxed optimum and a feasible solution thus automatically give a two-sided bound on the optimal objective. Clearly, the ideal outcome is for all inequalities to be met with equality. If in this case the relaxed and true optimal solutions coincide, we say that the relaxation is *exact* or *tight*. Bounding the size of the relaxation gap can provide substantial insight on how good an approximation a relaxation is, particularly if x' is a local minimum. While a relaxed optimal solution is in general not guaranteed to provide information about the location of true optimal solutions, a large body of empirical and theoretical work testifies to their quality as heuristic approximations.

A further useful feature of relaxations is that, if $x \in X, X \subset Y$, and

$$x \in \operatorname*{argmin}_{x \in Y} f(x),$$

then

$$x \in \operatorname*{argmin}_{x \in X} f(x).$$

In words, a relaxed optimal solution that is feasible for the original problem must be optimal for the original problem. This is easy to see by supposing its negation: if x is

---

[4] This itself is in fact a more general definition of a relaxation. Mathematically, the objective $h(x)$ and feasible set Y comprise such a relaxation if

$$\min_{x \in Y} h(x) \leq \min_{x \in X} f(x),$$

regardless of Y's relationship to X. For example, $h(x) = -\infty$ is a trivial relaxation for any Y. Both definitions are valid; we have arbitrarily presented the narrower one because it captures all relevant instances in this book.

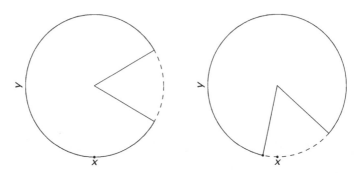

Fig. 2.8   Convex relaxations (dashed) of nonconvex sets (solid) with the objective $f(x, y) = y$. On the left, the relaxed minimum is feasible and thus optimal for the nonconvex problem. On the right, the relaxed minimum has a lower objective than the true minimum and is hence infeasible for the original problem.

relaxed optimal and feasible but not optimal for the original problem, then there exists an $x' \in X$ for which $f(x') < f(x)$. But then x' is also in Y, contradicting our assumption that x is relaxed optimal. We may conclude that a relaxed optimal solution must be either

- optimal for the original problem, or
- infeasible for the original problem.

This concept is illustrated in Figure 2.8. Note that a lower objective implies that a relaxed optimum is infeasible for the original problem, but the converse is not true: a relaxed optimum may be infeasible for the original problem but still have the same objective value as the true optimum.

2.3.1   Lift-and-project

This book employs a broad class of relaxations called *lift-and-project* methods. The basic approach is to "lift" nonconvex problems into a higher dimension where they are more nearly convex, which generally entails introducing new variables and constraints. The lifted feasible set is then made convex, often by simply removing nonconvex constraints. Finally, the solution of the lifted relaxation is "projected" back onto the original problem's variables. A basic trade-off exists between the accuracy and size of a relaxation. The application of the technique is illustrated through the following examples.

Example 2.7   *A small linear relaxation.* Consider the following optimization from Taylor and Hover [33]:

$$\text{minimize}_{x} \ x_1 (x_2 - 1)$$
$$\text{subject to} \ x_1 \geq 1, \quad x_2 \geq 2, \quad x_1 x_2 \leq 3.$$

One can verify from their Hessians that the objective and one constraint are nonconvex. Because this is a small problem, we can guess the optimal solution, $x_1 = 1$ and $x_2 = 2$. Suppose that we insist on solving this problem using convex optimization. Define a new variable, y, and substitute it for $x_1 x_2$. We now have a relaxation, which, because y is unbounded below, is trivially $-\infty$. We further constrain y by adding the redundant constraint $(x_1 - 1)(x_2 - 2) \geq 0$, which is called a *valid inequality*. Applying the same substitution to this constraint, we obtain the linear relaxation

$$\begin{aligned} \underset{x,y}{\text{minimize}} \quad & y - x_1 \\ \text{subject to} \quad & x_1 \geq 1, \quad x_2 \geq 2, \quad y \leq 3, \quad y - 2x_1 - x_2 + 2 \geq 0. \end{aligned}$$

We recover the solution by projecting the lifted solution back onto the original variables. In this example, this entails simply throwing out the new relaxation variable, y, and keeping x_1 and x_2. A relaxation is *tight* if the relaxed objective matches the original. A sufficient condition for tightness is the factorability of the lifted variables into the original ones at the optimal relaxed solution. The current problem is tight due to its small size, so that $y = x_1 x_2$. In general, we won't be so lucky, because if we were we would have a polynomial-time algorithm for many NP-hard problems!

The above is a linear relaxation obtained using the "reformulation-linearization technique," which is developed in generality in Sherali and Tuncbilek [34]; we will generically refer to this as lift-and-project. Its semidefinite generalization is known as the "moment-based relaxation" [35], the dual of which is called "sums-of-squares" [36, 37]. Although a dual version of the reformulation-linearization technique exists in theory [38], it has never found use as a numerical optimization tool. In each of these frameworks, one can construct more and more accurate relaxations by forming more valid inequalities and defining more variables. Eventually, all frameworks converge to the true optimal solution, albeit potentially with an exponential number of constraints [39]. Although the semidefinite hierarchies have been shown theoretically to converge faster than their linear counterparts [40], the superior speed of LP can make it a more pragmatic choice. Higher order and alternative conic relaxations are active areas of research, for example using hyperbolic programming [41], which subsumes all of the conic optimization classes covered here.

The following example is a special case of moment-based relaxations, which we refer to simply as the SD relaxation, and will play a key role in later chapters.

Example 2.8 *Semidefinite relaxation.* Consider the QCP:

$$\begin{aligned} \underset{x}{\text{minimize}} \quad & x^* Q x + c^* x \\ \text{subject to} \quad & x^* R_i x + r_i^* x + d_i \leq 0. \end{aligned}$$

If $Q \succeq 0$ and each $R_i \succeq 0$, the problem is convex and can be solved in polynomial-time by an interior point method. If this is not the case for either Q or an R_i, the problem is nonconvex and potentially very difficult to solve.

Introduce the Hermitian matrix X, and substitute X for the outer product xx^*. We can then write the above optimization equivalently as

$$\underset{x,X}{\text{minimize}} \quad \text{tr } QX + c^*x$$

$$\text{subject to} \quad \text{tr } R_iX + r_i^*x + d_i \leq 0$$

$$X = xx^*.$$

The last constraint is equivalent to the pair

$$\begin{bmatrix} X & x \\ x^* & 1 \end{bmatrix} \succeq 0 \quad \text{and} \quad \text{rank} \begin{bmatrix} X & x \\ x^* & 1 \end{bmatrix} = 1,$$

only the latter of which is nonconvex. Dropping the rank constraint yields the SD relaxation

$$\underset{x,X}{\text{minimize}} \quad \text{tr } QX + c^*x$$

$$\text{subject to} \quad \text{tr } R_iX + r_i^*x + d_i \leq 0$$

$$\begin{bmatrix} X & x \\ x^* & 1 \end{bmatrix} \succeq 0.$$

We can see from its Schur complement that the SD constraint is identical to $X \succeq xx^*$, so that the relaxation may be interpreted as weakening the equality $X = xx^*$. We have thus lifted a nonconvex problem with n variables to a convex SDP with $n(n+1)/2$ variables. If (X, x) and $\hat{x}$ are respectively the relaxed and true optimal solutions, then we are guaranteed that $\text{tr } QX + c^*x \leq \hat{x}^*Q\hat{x} + c^*\hat{x}$. The SD relaxation is in fact the dual of the famous *Shor relaxation* [42, 43], a precursor to many modern relaxation techniques.

In this book, we will only encounter special cases of the above QCP of the form

$$\underset{x,y}{\text{minimize}} \quad x^*Qx + c^*y$$

$$\text{subject to} \quad x^*R_ix + r_i^*y + d_i \leq 0.$$

For this problem, the SD relaxation takes on the slightly simpler form

$$\underset{X,y}{\text{minimize}} \quad \text{tr } QX + c^*y$$

$$\text{subject to} \quad \text{tr } R_iX + r_i^*y + d_i \leq 0$$

$$X \succeq 0.$$

Thus far we have described techniques for constructing linear and SD relaxations. Sometimes, however, a problem's "sweet spot" is in between, requiring something more expressive than LP but faster than SDP. This section concludes with a simple mechanism for relaxing SDPs to SOCPs, based on an observation from Kim and Kojima [44].

Example 2.9 *Relaxing SDPs to SOCPs.* Recall that a matrix is positive semidefinite if and only if all of its principal minors are nonnegative. A necessary condition for positive semidefiniteness is therefore the nonnegativity of all one-by-one and two-by-two principal minors, and replacing the former with the latter constitutes a relaxation. We refer to this as the *SOC relaxation*.

Mathematically, we express this as replacing the (real) SD constraint $X \succeq 0$ with the constraints

$$X_{ij}^2 \leq X_{ii}X_{jj}, \quad X_{ii} \geq 0, \quad \text{and} \quad X_{jj} \geq 0.$$

Nominally, there are $n(n-1)/2$ such constraint sets (one for each ij pair for which $i \neq j$), but we can often remove many of them because they contain "free" variables that are not in the objective or other constraints. $X_{ij}^2 \leq X_{ii}X_{jj}$ is a *hyperbolic constraint*, which may be written in standard SOC form as

$$\left\| \begin{bmatrix} 2X_{ij} \\ X_{ii} - X_{jj} \end{bmatrix} \right\| \leq X_{ii} + X_{jj}.$$

Many solvers can directly parse hyperbolic constraints but often require that the constraints $X_{ii} \geq 0$ and $X_{jj} \geq 0$ be stated explicitly even if they are implied elsewhere.

Nearly all of the SOC constraints encountered in this book are hyperbolic constraints; yet, as will be seen, the SD and SOC relaxations are all that are needed to "convexify" many essential problems in power systems.

The relaxation techniques presented in this section are applicable to any QCP. This actually goes well beyond our present needs. For instance, QCP's containment of integer programming enables us to apply lift-and-project methods to $x^2 = x$, which is equivalent to the binary constraint $x \in \{0, 1\}$. In fact, one of the original motivations for lift-and-project relaxations was to generate bounds in branch-and-bound routines by relaxing the binary constraint $x^2 = x$, cf. [45, 46]. We will not, however, apply lift-and-project in such scenarios but rather leave such integer constraints to the inner workings of algorithms.

Although we will only use lift-and-project methods to relax continuous constraints, it is worth mentioning their success in approximating NP-hard graph cut problems. We reproduce the seminal *Max-Cut* result of Goemans and Williamson [47] in Example 2.10 to illustrate the application of the SD relaxation. First, we must introduce some basic graph theory; the interested reader is referred to endnotes [48, 49] for more comprehensive treatments of graph theory.

2.3.2   *Detour:* graph theory

A *graph* is a set of *nodes* and *arcs*, respectively also called *vertices* and *edges*. Nodes have one index and are like points in space (or nodes in a power system), and arcs are like lines connecting the points and have two indices. If an arc exists between nodes i and j, it has a scalar weight, $w_{ij} > 0$. In an *undirected* graph, $w_{ij} = w_{ji}$, while in a directed graph this is not necessarily true. A *cut* is any set of arcs that when removed divides the nodes into two disconnected graphs. One can see that graph theory is implicit in much power system modeling.

A connected, undirected graph is radial if it has no loops or, equivalently, if there is exactly one path between every pair of nodes. Radial graphs are also referred to as acyclic graphs or trees.[5] A radial graph is shown in Figure 2.9. It turns out that radiality is an extremely useful property for a graph to possess. Graph theory is a source of numerous NP-hard and NP-complete combinatorial problems, cf. [8], many of which admit trivial or polynomial-time solutions if the graph is radial. Radial subgraphs are analogous to bases of *matroids*, which in some contexts guarantee the optimality of simple *greedy* algorithms [50]. More recently, it has been found that the radial structure of *poset causal* linear systems makes them amenable to optimal decentralized control [51]. This is just a small subset of the settings in which radiality is a powerful enabler, and future research is sure to uncover many more.

Returning to our present context, radiality is of both theoretical and practical importance in power systems as well. As we will see in Chapter 3, SOC relaxations of optimal power flow are provably exact under certain conditions in radial networks. The well-known linearized approximation is equivalent to an even simpler network flow approximation in radial networks. Distribution systems, which carry lower voltage electricity from substations to end users, are almost always operated radially for safety reasons. Transmission systems are usually very sparse, so that the mathematical benefits of radiality are only slightly degraded.

Fig. 2.9   A radial graph.

---

[5] This book uses the term "radial" because it is standard in power systems. However, "acyclic" and "tree" are more commonly used in graph theory and most other fields.

Example 2.10 *Max-Cut.* The NP-hard Max-Cut problem is to find the maximum weight cut of a graph with edge weights w_{ij} [52]. It can be written

$$\underset{x}{\text{maximize}} \sum_{ij} w_{ij} \left(1 - x_i x_j\right)$$

$$\text{subject to } x_i^2 = 1.$$

This is a difficult binary programming problem with a nonconvex objective (note, however, that the problem is solvable in polynomial-time if the graph is radial). Let us apply the SD relaxation from Example 2.8 by introducing the symmetric matrix X and substituting X_{ij} for $x_i x_j$ above. Max-Cut is then equivalently written as

$$\underset{x,X}{\text{maximize}} \sum_{ij} w_{ij} \left(1 - X_{ij}\right)$$

$$\text{subject to } X_{ii} = 1$$

$$X = xx^T.$$

As in Example 2.8, the SD relaxation of this problem is given by

$$\underset{X}{\text{maximize}} \sum_{ij} w_{ij} \left(1 - X_{ij}\right)$$

$$\text{subject to } X_{ii} = 1$$

$$X \succeq 0.$$

The main result of Goemans and Williamson [47] is that upon applying a randomized rounding procedure to the optimal X (the "projection" step), the expectation of the resulting cut weight is amazingly within a factor of 0.878 of the actual Max-Cut. This is, however, well outside the scope of this book.

2.3.3 *Preview:* How to use a relaxation

An important issue in this book is how to actually implement a solution from a relaxation. In some regards, the answer is just like any approximation: hope that it's close enough to the true optimal answer to achieve decent performance in reality. However, there is a specific issue we must address: when relaxed solutions aren't exact, they are infeasible. Why then is an infeasible relaxed solution better than a locally optimal but feasible solution? The answer is heavily context dependent. As a preview, let us summarize the implications of using relaxed solutions for each upcoming chapter.

Chapters 3 and 6: Relaxations for optimal power flow will sometimes be tight, in which case the relaxed solution is the true global optimum. It is shown in Section 3.3.1 that under certain conditions, this is guaranteed in radial networks. In general networks, easily verifiable conditions allow one to determine whether a particular instance is tight or not. When a relaxation is not tight, it often still provides a very good approximation. Coupled with a feasible solution, it provides a two-sided bound on the true optimum.

Chapter 4: The problems in this chapter inherit the characteristics of the underlying power flow relaxation. In reconfiguration, the network is constrained to be radial, so that relaxed solutions are usually feasible and thus optimal due to the results of Section 3.3.1. Even if the power flow solution is not feasible, a possibly suboptimal but feasible radial configuration is guaranteed. In unit commitment, relaxed power flow constraints can lead to under-scheduling of resources; however, status quo unit commitment models employ far simpler power flow models that admit solutions that are significantly further from feasibility.

Chapter 5: Component placement problems often start feasible, and the goal is to place new resources to further improve performance. In this case, relaxed power flows may be infeasible, but the resulting component placement plan is a feasible approximation. Transmission planning approximations will generally be infeasible but effectively allow one to prune the search space, making subsequent design stages more focused and efficient.

Beyond each problem's particularities, there are two salient characteristics that make convex relaxations useful in any environment: speed and reliability. Because they are amenable to polynomial-time algorithms, robust, efficient algorithms are readily applicable. They are reliable because they are consistent: whereas locally optimal solutions of nonconvex formulations may depend on an algorithm's starting point, convex relaxations only have global optima.

2.4   Classical optimization versus metaheuristics

In the 1990s, various so-called metaheuristic algorithms came to prominence in power systems for their alleged ability to solve problems that were troublesome for classical optimization algorithms. These include genetic algorithms, particle swarms, tabu search, evolutionary algorithms, ant colony optimization, harmony search, bacterial foraging; the list goes on. At the time of writing, virtually every metaheuristic algorithm has been applied to every optimization problem in power systems.

There is obvious appeal to metaheuristic algorithms. Most function by searching multiple portions of the solution space in parallel, with each search's next move typically determined by information from the other searches and a random element. They are based on simple, intuitive rules that make them easy to understand and (initially) program. They make no assumptions about problem structure, so that they are applicable to nearly everything. They are often accompanied by aggressive performance claims such as immunity to nonconvexity and noise, and they almost always have very exciting names.

Unfortunately, most such claims are theoretically unsubstantiated, and the broad applicability comes at the expense of excellence within any particular problem class. Indeed, one would never use a metaheuristic to solve a convex cone program like those in Section 2.2 for the same reason one would never use the bisection method to find the root of a scalar quadratic equation. Furthermore, metaheuristics have found little to no usage in the power industry despite the amount of literature devoted to them. Some of the technical reasons are made precise below.

Convergence: Classical optimization algorithms provably converge to a minimum and tell you when and how precisely they have converged. Most metaheuristics have been shown to converge in numerical experiments, but theoretical guarantees usually do not exist, and in some cases contrary evidence has been found [53].

Complexity: The scaling of convergence rate with problem size is known for most classical optimization techniques, as discussed in Section 2.1. No such analyses exist for metaheuristics because convergence itself is not established.

Implementation: Classical optimization algorithms are generally available in professional-grade software packages, so that one only has to program the model. Metaheuristic algorithms are less standardized and have yet to see professional-grade software implementations. Consequently, metaheuristic approaches usually entail programming both model and algorithm, as well as problem-specific parameter tuning.

Classical optimization algorithms have better convergence properties than metaheuristics because they make extensive use of structural information like gradients, Hessians, and sparsity. For the optimization classes in Section 2.2, this structure is specified just by writing the problem in standard form.

In fairness, a metaheuristic algorithm *could* perform better than a classical algorithm like Newton's method on a badly nonconvex problem. It is rather the dangerous philosophy of ignoring a problem's mathematical structure that will lead one to far less effective solutions and which has often led to easy problems being labeled difficult. As will be seen over the course of this book, many optimization problems in power systems possess rich and elegant structure, making them highly amenable to rigorous and powerful classical optimization algorithms.

2.5 Power system modeling

In large-scale optimization, detailed physical modeling necessarily takes a backseat to algorithmic tractability. In power systems, this means abandoning many significant details to reach models amenable to the algorithms discussed earlier in this chapter. Skipping ahead, the model we employ is a single-phase, steady-state approximation in which the voltage, current, and power across a line connecting two nodes satisfy

$$p_{ij} + iq_{ij} = v_i i_{ij}^* \tag{2.3}$$

$$= v_i \left(v_i^* - v_j^* \right) y_{ij}^*, \tag{2.4}$$

where y_{ij} is the line's complex admittance. We furthermore work with dimensionless *per unit* quantities, so that all voltage magnitudes are nominally close to one. Finally, we assume balanced operation, enabling the entire system to be analyzed through a

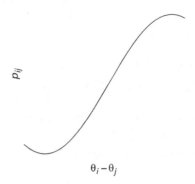

Fig. 2.10   The nonconvex equality constraint relating voltage and power, projected onto voltage angle
difference and real power flow variables, $\theta_i - \theta_j$ and p_{ij}.

single phase. This section summarizes some of the technical details leading up to this
model. The reader is referred to endnotes [54–57] for comprehensive expositions of
power system modeling.

Before proceeding, however, an immediate question we must answer is: why power?
Voltage and current share a far simpler relationship through Ohm's and Kirchhoff's
laws than power's quadratic dependence on voltage. Indeed, (2.4) is a quadratic equality
constraint and therefore nonconvex, as illustrated in Figure 2.10. So why not model
power systems in terms of voltage and current if the resulting model would be far
more tractable? The reason is that voltage or current alone is not useful. Loads need
power to do work like lighting, temperature regulation, and manufacturing. Hence, a
load cannot be specified in terms of voltage without also making some assumption
about current and vice versa, effectively specifying their power need. Analogously,
power plants burn fuel or harvest energy from water, wind, and sunlight, and they
are primarily paid for the power they produce. For these reasons, modeling power
systems in terms of power is unfortunately an inescapable reality of power system
engineering.

2.5.1   Voltage, current, and power in steady-state

We write alternating voltage and current as the sinusoids

$$V = \overline{V}\cos(\omega t + \theta) \quad \text{and} \quad I = \overline{I}\cos(\omega t + \theta - \phi),$$

where $\overline{V}$ and $\overline{I}$ are their magnitudes. ω is frequency in radians; in North America,
$\omega = 120\pi$, and in Europe $\omega = 100\pi$. Our first assumption is that voltage and current can indeed be written as above, when in fact they are never true sinusoids and
exist alongside higher frequency harmonic signals. This is a well-founded approximation because higher harmonics are usually small in magnitude and voltage and current

are very nearly sinusoidal. Currently, this is because most electric power is supplied by synchronous generators, which are operated tightly around a nominal frequency.[6]

We next assume that voltage and current are in steady-state, which is to say that ω, $\overline{V}, \overline{I}, \theta$, and ϕ do not vary with time. Of course, this is never exactly true, but there are many scenarios for which the effects of minor changes in these quantities are insignificant relative to other physical phenomena. This enables phasor notation, which is used exclusively throughout the remaining chapters.

The product of voltage and current is power, which, using basic trigonometric identities, can be written

$$
\begin{aligned}
P &= VI \\
&= \overline{VI} \cos(\omega t + \theta)\cos(\omega t + \theta - \phi) \\
&= \frac{\overline{VI}}{2} \cos(\phi)\,(1 + \cos(2(\omega t + \theta))) + \frac{\overline{VI}}{2} \sin(\phi)\sin(2(\omega t + \theta)).
\end{aligned}
\tag{2.5}
$$

This is referred to as the *instantaneous power*. Notice that each term has a constant multiplying a time-varying quantity, which has twice the frequency of voltage and current. Over a period of length π/ω, the first term's average is this constant portion, $\overline{VI} \cos(\phi)/2$, and the second's is zero. The constant coefficient of the first term is called *real power* and is responsible for doing work like heating a building or spinning a motor. The constant coefficient of the second term is known as *reactive power*. Although reactive power does not deliver energy, it plays an important role in AC power systems and electrical loads, such as sustaining the magnetic fields of induction motors and transformers.

We can write voltage and current

$$
V = \operatorname{Re} \overline{V} e^{j\theta} e^{j\omega t} \quad \text{and} \quad I = \operatorname{Re} \overline{I} e^{j(\theta - \phi)} e^{j\omega t},
$$

which motivates the definition of the voltage and current steady-state phasors,

$$
v = \frac{\overline{V}}{\sqrt{2}} e^{j\theta} \quad \text{and} \quad i = \frac{\overline{I}}{\sqrt{2}} e^{j(\theta - \phi)}.
$$

Note that $\overline{V}/\sqrt{2}$ and $\overline{I}/\sqrt{2}$ are the respective root mean square magnitudes of V and I. Observe that

$$
vi^* = \frac{\overline{VI}}{2}(\cos(\phi) + i\sin(\phi)),
$$

the real and imaginary parts of which respectively match the constant portions of the terms in (2.5). We hence define the steady-state real and reactive power phasors

$$
\begin{aligned}
p &= \frac{\overline{VI}}{2} \cos(\phi) \\
&= \operatorname{Re} vi^*
\end{aligned}
$$

---

[6] As renewable sources like wind and solar replace synchronous generators and direct current applications such as high-voltage transmission grow, the prevalence of this paradigm may recede.

and

$$q = \frac{\overline{VI}}{2}\sin(\phi)$$
$$= \operatorname{Im} vi^{*}.$$

From these we recover (2.3). The quantity $\cos(\phi)$ is known as the *power factor* and indicates how much of the power is real versus reactive.

Now suppose power is flowing from a source at a node, labeled i, through a power line to a load at another node labeled j. The line is described by the complex impedance

$$z_{ij} = r_{ij} + ix_{ij}.$$

The real part is the resistance and is always positive and causes real power losses through heating. The imaginary part is the reactance and is a function of the line's inductance, L_{ij}, capacitance, C_{ij}, and frequency,

$$x_{ij} = \omega L_{ij} - \frac{1}{\omega C_{ij}}.$$

If the line is inductive, $x_{ij} > 0$ and the line consumes reactive power. If it is capacitive, $x_{ij} < 0$ and the line produces reactive power.

Admittance is the reciprocal of impedance, denoted

$$y_{ij} = \frac{1}{r_{ij} + ix_{ij}}$$
$$= \frac{r_{ij}}{r_{ij}^2 + x_{ij}^2} - i\frac{x_{ij}}{r_{ij}^2 + x_{ij}^2}$$
$$= g_{ij} - ib_{ij}.$$

The real part is now referred to as conductance and the imaginary part susceptance.[7] Usually, $b_{ij} > 0$ due to the inductive nature of power lines. These and the other quantities in this section are illustrated in Figure 2.11.

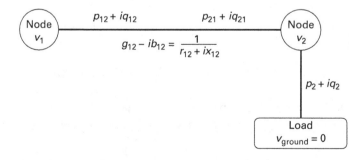

Fig. 2.11   Power flowing from node one to a load at node two.

---

[7] We have neglected *shunt* admittances describing each line's interaction with ground because they increase the amount of necessary notation and their effects are usually small. However, they can be straightforwardly added to all models in this book.

Ohm's law says that the current between the nodes is related to the nodal voltages as

$$i_{ij} = (v_i - v_j)\, y_{ij}.$$

The power through the line is obtained by multiplying the conjugate of the current by the voltage rise between ground and node i or, alternatively, the voltage drop between node i and the ground on the other side of the load at node j. This yields

$$p_{ij} + iq_{ij} = v_i\left(v_i^* - v_j^*\right)y_{ij}^*.$$

Note that this is the power at node i going to j. The power arriving at j from i is this quantity minus the complex power losses along the line. The power departing j for i is simply the negative of the arriving power and is thus

$$p_{ji} + iq_{ji} = |v_i - v_j|^2\, y_{ij}^* - p_{ij} - iq_{ij} \tag{2.6}$$

$$= v_j\left(v_j^* - v_i^*\right)y_{ij}^*.$$

Notice that as the magnitudes of v_i and v_j increase at the same rate, the loss term in the first line remains constant, but the total power flow as written in the second line increases. It is for this reason that electric power is transmitted at far higher voltages than those at which it is generated and consumed.

2.5.2 Balanced three-phase operation

Most power lines have multiple phases wherein separate wires carry power in parallel. The most common configuration is to have three *balanced* phases, in which each line is identical except that their phases are evenly staggered such that

$$V_a = \overline{V}\cos(\omega t + \theta)$$
$$V_b = \overline{V}\cos\left(\omega t + \theta + \frac{2\pi}{3}\right)$$
$$V_c = \overline{V}\cos\left(\omega t + \theta - \frac{2\pi}{3}\right)$$

and

$$I_a = \overline{I}\cos(\omega t + \theta - \phi)$$
$$I_b = \overline{I}\cos\left(\omega t + \theta + \frac{2\pi}{3} - \phi\right)$$
$$I_c = \overline{I}\cos\left(\omega t + \theta - \frac{2\pi}{3} - \phi\right).$$

We discuss other configurations in Example 2.11. The currents satisfy

$$I_a + I_b + I_c = 0. \tag{2.7}$$

This leads to considerable savings by theoretically eliminating the return current and hence the need for a return conductor (in practice, the phases are never perfectly balanced, and so a *neutral* line is often included). The second major benefit of having

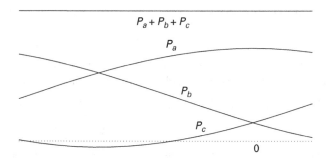

Fig. 2.12   Three balanced phases serving constant total instantaneous power.

three balanced phases is that the total instantaneous power is constant, as seen below by summing (2.5) over each phase:

$$P = P_a + P_b + P_c \tag{2.8}$$

$$= \frac{\overline{VI}}{2} \cos(\phi) \left[3 + \cos(2(\omega t + \theta)) + \cos\left(2\left(\omega t + \theta + \frac{2\pi}{3}\right)\right) \right.$$

$$\left. + \cos\left(2\left(\omega t + \theta - \frac{2\pi}{3}\right)\right) \right] + \frac{\overline{VI}}{2} \sin(\phi) \left[\sin(2(\omega t + \theta)) \right.$$

$$\left. + \sin\left(2\left(\omega t + \theta + \frac{2\pi}{3}\right)\right) + \sin\left(2\left(\omega t + \theta - \frac{2\pi}{3}\right)\right) \right] \tag{2.9}$$

$$= \frac{3\overline{VI}}{2} \cos(\phi). \tag{2.10}$$

The arguments of the three cosine terms and three sine terms are respectively balanced and therefore add to zero, as in (2.7). From this it is clear that the total instantaneous power is just three times the steady-state or real power in a single phase. This concept is illustrated in Figure 2.12. Constant power is particularly beneficial to three-phase generators and loads. For instance, the rotor of a generator or motor will experience less stress and hence require less maintenance under constant torque. Many loads, however, simply receive power from a single phase, and balanced system operation is attained by evenly apportioning the loads over the phases.

It can be shown that if certain assumptions are met in addition to three balanced phases, any three-phase network can be decomposed into three identical single-phase networks, allowing the entire system to be represented by a single phase. This is a standard practice in power systems, which we subsequently adopt for the remainder of this book.

Example 2.11   *Balanced n-phase operation.* This example considers power systems with more than three phases. As in the three-phase case, voltage and current are given by

$$V_k = \overline{V}\cos(\omega t + \theta_k) \quad \text{and} \quad I_k = \overline{I}\cos(\omega t + \theta_k - \phi)$$

for $k = 0, \ldots, n - 1$. The total instantaneous power delivered is

$$P = \sum_k V_k I_k.$$

We are interested in configurations that, like three-phase operation, result in constant total instantaneous power and zero return current,

$$\sum_k I_k = 0.$$

Recalling (2.7) and (2.8)-(2.10), we can express these requirements as

$$\sum_k e^{i\theta_k} = 0 \tag{2.11}$$

$$\sum_k e^{i2\theta_k} = 0 \tag{2.12}$$

A sufficient condition for (2.11) and (2.12) is for both the angles and double angles to be balanced sets.

With n phases, the balanced set of angles is

$$\theta_k = \frac{2\pi k}{n}, \quad k = 0, \ldots, n - 1. \tag{2.13}$$

We can immediately disqualify $n = 2$ because any choice of θ satisfying (2.11) results in $e^{i2\theta_1} = e^{i2\theta_2} \neq 0$ and thus cannot produce constant instantaneous power. For $n \geq 3$, (2.13) automatically satisfies (2.11). If n is even, (2.12) can be written

$$\sum_{k=0}^{n-1} e^{i2\theta_k} = \sum_{k=0}^{\frac{n}{2}-1} e^{\frac{i4\pi k}{n}} + \sum_{k=\frac{n}{2}}^{n-1} e^{\frac{i4\pi k}{n}}$$

$$= \sum_{k=0}^{\frac{n}{2}-1} e^{\frac{i4\pi k}{n}} + \sum_{k=0}^{\frac{n}{2}-1} e^{\frac{i4\pi k}{n}} e^{i2\pi}$$

$$= 2\sum_{k=0}^{\frac{n}{2}-1} e^{\frac{i4\pi k}{n}}$$

$$= 0.$$

Observe that the last equality holds because the angles

$$\theta_k' = \frac{4\pi k}{n}, \quad k = 0, \ldots, \frac{n}{2} - 1$$

are also spaced evenly about the origin.

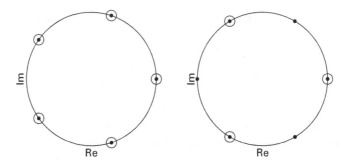

Fig. 2.13   Balanced five-phase (left) and six-phase (right) configurations both satisfy (2.11) and (2.12).
Voltage angles are dots and power angles circles viewed on the complex plane. With five phases,
the voltage and power angles are identical, and with six, the power angles appear twice at the
doubled voltage angles.

Now suppose that n is odd. Then (2.12) is given by

$$\sum_{k=0}^{n-1} e^{i2\theta_k} = \sum_{k=0}^{\frac{n-1}{2}} e^{\frac{i4\pi k}{n}} + \sum_{k=\frac{n+1}{2}}^{n-1} e^{\frac{i4\pi k}{n}}$$

$$= \sum_{k=0}^{\frac{n-1}{2}} e^{\frac{i4\pi k}{n}} + \sum_{k=0}^{\frac{n-3}{2}} e^{\frac{i4\pi(k+1/2)}{n}} e^{i2\pi}$$

$$= \sum_{k=0}^{n-1} e^{\frac{i2\pi k}{n}}$$

$$= 0.$$

Here, the set of power angles, $2\theta_k$, is identical to the set of current angles, θ_k, and is
thus balanced as well. These concepts are illustrated in Figure 2.13.

We have thus shown that as long as there are three or more balanced phases, zero
return current and constant instantaneous power are guaranteed. Of course, three
phases is the undisputed standard for long-distance AC power transmission because it
is the simplest configuration with these properties.

2.5.3   Generator and load modeling

In power systems, groups of components collectively referred to as generators and loads
are located at nodes throughout the network. To manageably incorporate these components into system-level models, it is standard to use highly simplified descriptions in
which each generator and load is parameterized by its real power p, reactive power q,
and complex voltage v. Typically, these quantities are subject to *box constraints* of the
form

$$\underline{p}_i \le p_i \le \overline{p}_i$$
$$\underline{q}_i \le q_i \le \overline{q}_i$$
$$\underline{v}_i \le |v_i| \le \overline{v}_i,$$

where the last constraint is on voltage magnitude.

Generators

Generators convert energy in various forms to electrical energy, which they sell to power system operators and loads. Much of power system operation is structured around the cost of producing power, which depends on many factors like the cost of fuel (e.g., coal, oil, uranium), how efficiently a fuel's energy can be transformed to electrical power, and the maintenance costs associated with normal and extreme operating conditions [58].

At the system level, the cost of producing electric power is condensed into a single function known by various names such as the *fuel curve*, *production cost curve*, or just *cost curve* for short. Most of this book will assume that this function is a convex polynomial, for example the quadratic

$$f_i(p_i) = a_i p_i^2 + b_i p_i + c_i, \quad 0 \le p_i \le \overline{p}_i.$$

Special types of nonconvex costs are discussed in Section 3.5.3. When $f_i(p_i)$ is submitted by generators into an electricity market, it is called a *supply function*. Often, supply functions describe the amount of power they are willing to sell at a certain price, i.e., $g_i(\lambda) = f_i^{-1}(p_i \lambda)$, where λ is the price of power.

The sum of the generators' cost curves is the objective of many of the power system optimizations covered in this book. Note that because reactive power carries no energy, it is not associated with any significant costs. However, reactive power can limit real power production leading to opportunity costs, which are discussed in Section 3.5.2 and Example 6.6.

Loads

With the exception of large components like factory induction motors, most loads are inherently uncertain in the eyes of system operators, simply because there are far too many to keep track of and much of their energy usage is private. Consequently, approximate load models of load aggregations are often used over models derived from first principles.

A common parameterization makes real and reactive power quadratic functions of voltage magnitude:

$$p_i = p_i^P + p_i^I |v_i| + p_i^Z |v_i|^2$$
$$q_i = q_i^P + q_i^I |v_i| + q_i^Z |v_i|^2$$

The constant, linear, and squared terms correspond to constant power (P), current (I), and impedance (Z), which has led this to be referred to as the *ZIP* model [59].

Unfortunately, even the minimal ZIP model makes most optimization routines nonconvex, so that purely constant power load models are almost exclusively encountered in the academic literature. Example 3.4 shows how to approximately incorporate ZIP models into relaxations without sacrificing convexity.

Traditionally, the idea of a load preference was nonsensical because demand was expected to be met exactly, and there was no benefit in surplus. Now in some modern scenarios, loads may submit demand curves similar to supply functions, which reflect their aggregate active participation in power markets, for example, through demand response programs, real-time pricing, or unreliable service contracts, cf. [60–62].

2.5.4   The per unit system

The per unit system is a method of normalizing voltage, current, power, and impedance according to their dimensional interrelationships. After this section, we will implicitly assume that all quantities are per unit quantities normalized about nominal levels. This serves to streamline many expository aspects, such as when we fix all voltage magnitudes at one per unit in Section 3.2.1.

To proceed, choose real base values for two quantities, for instance, voltage, v_b, and impedance, z_b, for a portion of the network. For example, at the high voltage transmission level, one might use $v_b = 765,000$ volts. The per unit voltage and impedance are then

$$v_{pu} = \frac{v}{v_b} \quad \text{and} \quad z_{pu} = \frac{z}{z_b}.$$

The corresponding current and power bases are obtained by Ohm's law,

$$i_b = \frac{v_b}{z_b} \quad \text{and} \quad s_b = \frac{v_b^2}{z_b}.$$

Note that any pair of base quantities could be used with no functional change. The resulting dimensionless per unit quantities satisfy all of the original physical relationships, and actual quantities are straightforwardly recovered by multiplying by base values. By choosing appropriate base quantities at different network locations, components at different voltage levels can be conveniently modeled together. This has the benefit of adding intuition to heterogenous models, improving the numerical conditioning of computations, and removing transformers from large models, as demonstrated in Example 2.12.

Example 2.12 *Per unit normalization with an ideal transformer.* Transformers vary the ratio of current to voltage, enabling low-voltage, high-current generation and consumption and high-voltage, low-current transmission at fixed power levels. Because low-voltage, long-distance transmission is impractical due to losses, as discussed in Section 2.5.1, transformers are essential components of large power systems. The vast majority of power systems are AC because transformers do not work in DC systems.

A single-phase, ideal transformer has a *primary* side through which power enters at a certain voltage and current and a *secondary* side from which the same quantity of power exits with a different voltage and current. The change in the ratio of voltage and current is determined by the number of *winding turns* on the primary side, n_1, and on the secondary side, n_2, as

$$\frac{v_1}{v_2} = \frac{n_1}{n_2} \quad \text{and} \quad \frac{i_1}{i_2} = \frac{n_2}{n_1}.$$

It is easy to see that the power on the primary side is equal to that on the secondary side, i.e., $s_1 = v_1 i_1^* = v_2 i_2^* = s_2$.

If we choose our per unit base quantities such that

$$\frac{v_{1,b}}{v_{2,b}} = \frac{n_1}{n_2} \quad \text{and} \quad s_{1,b} = s_{2,b},$$

we have $v_{1,pu} = v_{2,pu}$, $i_{1,pu} = i_{2,pu}$ and $s_{1,pu} = s_{2,pu}$. The per unit system can thus be used to eliminate transformers from power flow models. Throughout this book, it is assumed that all quantities have been normalized in this fashion.

The swing bus

When *solving* for the power flow in a network (called *load flow*, discussed in Section 3.4), one determines all complex voltages given that nodes are specified in terms of p and $|v|$ (generators) or p and q (loads). This alone, however, is not enough to specify a single solution because there are an infinite number of feasible voltage angles. This is remedied by fixing an arbitrary *swing bus* at $v_s = 1$ per unit.

The convention is usually unnecessary in our context because optimization entails searching through a continuous set of feasible voltages. Moreover, when voltage limits are specified in a transmission network, fixing a particular node's voltage at one can make potentially good solutions infeasible. As such, the swing bus does not feature in most formulations in this book.

2.6   Summary

We have now assembled the basic tools necessary to formulate the power system optimization problems in this book. In doing so, we have highlighted recent theoretical developments that underlie some of the more modern material, such as generalizations of LP that admit polynomial-time algorithms and lift-and-project relaxations. We have simultaneously identified avenues for future theoretical research that will surely bring further enhancements to computation in power systems, some of which are listed below:

- While SDPs are solvable in polynomial-time, SDP algorithms and software implementations are not as mature as LP and SOCP algorithms. The challenge lies in leveraging *sparsity*, which is automatically exploited by LP and SOCP algorithms but requires additional treatment in SDP [63].
- Over the course of this book, we will encounter a variety of problems with discrete elements, particularly in Chapters 4 and 5. Presently, only MILPs are efficiently solvable, but MISOCP algorithms are rapidly improving [30–32]. While one can

never expect algorithms for MISOCP and MISDP to be as fast as those for MILP, there is considerable room for development before they can be said to have attained the same maturity.

- Lift-and-project relaxations are relatively mysterious in that very few theoretical results exist for predicting their quality, for example, by bounding the gap between relaxed and optimal solutions as for Max-Cut in endnote [47]. The next chapter will show that some relaxations are exact under certain conditions, but we are otherwise unable to assess how accurate they are without doing numerical tests. New theoretical tools for relaxations would therefore greatly increase their relevance; however, whereas progress on the previous two points is fairly likely, there is no clear route to obtaining such general theoretical results for relaxations.

Improvements in any of these basic areas will advance power systems and many other fields. Over the course of this book, we will see that there are many such fundamental and application-driven sides of power systems where there is room for research and development.

Problems

2.1   Let $C \succ 0$. Analytically characterize the optimal solution of the following equality constrained convex quadratic program:

$$\underset{x}{\text{minimize}} \ \ x^* C x$$

$$\text{subject to} \ \ Ax = b.$$

2.2   Write the following as a standard form SOC constraint:

$$x^2 - (x - 5)y - yz + 3(z - 5)^2 \le 1 + x.$$

2.3   Recall the SD formulation of eigenvalue optimization in Example 2.2. Suppose now that we want to minimize the second largest eigenvalue of the matrix $A(x)$ and that the eigenvector of the largest eigenvalue, v, does not depend on x. Write this as an SDP.

2.4   Consider the following optimization:

$$\underset{x}{\text{minimize}} \ \ c^T x$$

$$\text{subject to} \ \ x_i x_{i+2} = 0$$

$$x_i \ne 0 \text{ for at most three } i\text{'s}$$

$$Ax \le b.$$

Use integer constraints to write this as an MILP.

2.5   Define the function

$$f(x) = c_i^T x \quad \text{if} \quad a_i \le x_1 \le a_{i+1},$$

where each a_i is given. Write the following optimization as an MILP:

$$\underset{x}{\text{minimize}} \ \ f(x)$$

$$\text{subject to} \ \ Ax \le b.$$

2.6 In graph theory, a path is a sequence of adjacent edges between two nodes. Show that in an undirected radial graph, there is exactly one path between each pair of nodes.

2.7 Consider an undirected graph with edge weights w_{ij}. The graph's Laplacian matrix is defined as

$$L_{ij} = \begin{cases} -w_{ij} & \text{if } i \neq j \\ \sum_j w_{ij} & \text{if } i = j \end{cases}.$$

Show that Laplacian matrix is positive semidefinite and has rank $n - 1$ (one less than the dimension).

2.8 Use the SD relaxation of Example 2.8 to obtain a (continuous) relaxation for the following MILP:

$$\begin{aligned} \underset{x}{\text{minimize}} \quad & c^T x \\ \text{subject to} \quad & Ax \leq b \\ & x_i \in \{0, 1\}. \end{aligned}$$

2.9 Suppose $A_{ij} = 0$ if i times j is even. Use the SOC relaxation to relax the following SDP to an SOCP:

$$\begin{aligned} \underset{X}{\text{minimize}} \quad & \text{tr } X \\ \text{subject to} \quad & \text{tr } AX \leq \sum_i b_i X_{ii} \\ & X \succeq 0. \end{aligned}$$

How many fewer variables does the resulting SOCP need than the original SDP?

2.10 Consider an n-phase power system as in Example 2.11, i.e., with

$$V_k = \bar{V} \cos(\omega t + \theta_k) \quad \text{and} \quad I_k = \bar{I} \cos(\omega t + \theta_k - \phi)$$

for $i = 1, \ldots, n$ and

$$P = \sum_k V_k I_k.$$

Construct a set of unevenly spaced θ_k for which the return current, $\sum_k I_k$, is zero and the total instantaneous power, P, is constant.

References

[1] S. Boyd and L. Vandenberghe, *Convex Optimization*. New York: Cambridge University Press, 2004.

[2] D. P. Bertsekas, *Nonlinear Programming*. Athena Scientific, 1999.

[3] M. Bazaraa, H. Sherali, and C. Shetty, *Nonlinear Programming Theory and Algorithms*. New York: John Wiley, 1993.

[4] Y. Nesterov and A. Nemirovski, *Interior Point Polynomial Methods in Convex Programming*. PSIAM Studies in Applied Mathematics, 1994, vol. 13.

[5] J. Nocedal and S. J. Wright, *Numerical Optimization*. Springer, 2000.

[6] R. T. Rockafellar, *Convex Analysis*, ser. Princeton Landmarks in Mathematics and Physics. Princeton University Press, 1996.

[7] D. Bertsekas, A. Nedić, and A. Ozdaglar, *Convex Analysis and Optimization*, ser. Athena Scientific Optimization and Computation Series. Athena Scientific, 2003.

[8] M. R. Garey and D. S. Johnson, *Computers and Intractability: A Guide to the Theory of NP-Completeness (Series of Books in the Mathematical Sciences)*. W. H. Freeman, Jan. 1979.

[9] C. M. Papadimitriou, *Computational Complexity*. Reading, MA: Addison-Wesley, 1994.

[10] S. Arora and B. Barak, *Computational Complexity: A Modern Approach*. Cambridge University Press, 2009.

[11] D. Bertsimas and J. N. Tsitsiklis, *Introduction to Linear Optimization*. Athena Scientific, 1997.

[12] G. B. Dantzig, *Linear Programming and Extensions*. Princeton University Press, 1998.

[13] C. H. Papadimitriou and K. Steiglitz, *Combinatorial Optimization: Algorithms and Complexity*. Dover Publications, 1998.

[14] M. S. Bazaraa, J. J. Jarvis, and H. D. Sherali, *Linear Programming and Network Flows*. Wiley-Interscience, 2004.

[15] P. M. Pardalos and S. A. Vavasis, "Quadratic programming with one negative eigenvalue is NP-hard," *Journal of Global Optimization*, vol. 1, no. 1, pp. 15–22, 1991.

[16] N. Karmarkar, "A new polynomial-time algorithm for linear programming," in *Proceedings of the Sixteenth Annual ACM Symposium on Theory of Computing*, ser. STOC '84. New York: ACM, 1984, pp. 302–311.

[17] M. S. Lobo, L. Vandenberghe, S. Boyd, and H. Lebret, "Applications of second-order cone programming," *Linear Algebra and Its Applications*, vol. 284, pp. 193–228, Nov. 1998.

[18] F. Alizadeh and D. Goldfarb, "Second-order cone programming," *Mathematical Programming*, vol. 95, pp. 3–51, 2003.

[19] L. Vandenberghe and S. Boyd, "Semidefinite programming," *SIAM Review*, vol. 38, no. 1, pp. 49–95, 1996.

[20] R. A. Horn and C. R. Johnson, *Matrix Analysis*. Cambridge University Press, 1990.

[21] S. Boyd, L. El Ghaoui, E. Feron, and V. Balakrishnan, *Linear Matrix Inequalities in System and Control Theory*, ser. Philadelphia: SIAM Studies in Applied Mathematics, 1994, vol. 15.

[22] M. P. Bendsøe and O. Sigmund, *Topology Optimization: Theory, Methods and Applications*, 2nd ed. Springer, 2003.

[23] S. Boyd, P. Diaconis, and L. Xiao, "Fastest mixing Markov chain on a graph," *SIAM Review*, vol. 46, no. 4, pp. 667–689, 2004.

[24] L. Xiao and S. Boyd, "Fast linear iterations for distributed averaging," *Systems & Control Letters*, vol. 53, no. 1, pp. 65–78, 2004.

[25] L. A. Wolsey, *Integer Programming*. Wiley-Interscience, 1998.

[26] A. Schrijver, *Theory of Linear and Integer Programming*. John Wiley & Sons, June 1998.

[27] L. A. Wolsey and G. L. Nemhauser, *Integer and Combinatorial Optimization*. Wiley-Interscience, 1999.

[28] E. Balas, "Disjunctive programming and a hierarchy of relaxations for discrete optimization problems," *SIAM Journal on Algebraic Discrete Methods*, vol. 6, no. 3, pp. 466–486, 1985.

[29] I. E. Grossmann, "Review of nonlinear mixed-integer and disjunctive programming techniques," *Optimization and Engineering*, vol. 3, no. 3, pp. 227–252, 2002.

[30] J. P. Vielma, S. Ahmed, and G. L. Nemhauser, "A lifted linear programming branch-and-bound algorithm for mixed-integer conic quadratic programs," *INFORMS Journal on Computing*, vol. 20, pp. 438–450, July 2008.

[31] S. Drewes, "Mixed integer second order cone programming," Ph.D. dissertation, Technischen Universität Darmstadt, Department of Mathematics, 2009.

[32] A. Atamtürk and V. Narayanan, "Conic mixed-integer rounding cuts," *Mathematical Programming*, vol. 122, pp. 1–20, 2010.

[33] J. A. Taylor and F. S. Hover, "Linear relaxations for transmission system planning," *IEEE Transactions on Power Systems*, vol. 26, no. 4, pp. 2533–2538, Nov. 2011.

[34] H. D. Sherali and C. H. Tuncbilek, "A global optimization algorithm for polynomial programming problems using a reformulation-linearization technique," *Journal of Global Optimization*, vol. 2, pp. 101–112, 1992.

[35] J. B. Lasserre, "Global optimization with polynomials and the problem of moments," *SIAM Journal on Optimization*, vol. 11, no. 3, pp. 796–817, 2000.

[36] P. A. Parrilo, "Structured semidefinite programs and semialgebraic geometry methods in robustness and optimization," Ph.D. dissertation, California Institute of Technology, Pasadena, CA, 2000.

[37] ——, "Semidefinite programming relaxations for semialgebraic problems," *Mathematical Programming*, vol. 96, no. 2, pp. 293–320, 2003.

[38] K. Schmüdgen, "The K-moment problem for compact semi-algebraic sets," *Mathematische Annalen*, vol. 289, no. 1, pp. 203–206, March 1991.

[39] J. B. Lasserre, "Polynomial programming: LP-relaxations also converge," *SIAM Journal on Optimization*, vol. 15, no. 2, pp. 383–393, 2005.

[40] ——, "Semidefinite programming vs. LP relaxations for polynomial programming," *Mathematics of Operations Research*, vol. 27, no. 2, pp. 347–360, 2002.

[41] J. Renegar, "Hyperbolic programs, and their derivative relaxations," *Foundations of Computational Mathematics*, vol. 6, no. 1, pp. 59–79, 2006.

[42] N. Z. Shor, "Quadratic optimization problems," *Soviet Journal Computer and Systems Sciences*, vol. 25, pp. 1–11, 1987.

[43] ——, "Dual quadratic estimates in polynomial and boolean programming," *Annals of Operations Research*, vol. 25, pp. 163–168, 1990.

[44] S. Kim and M. Kojima, "Exact solutions of some nonconvex quadratic optimization problems via SDP and SOCP relaxations," *Computational Optimization and Applications*, vol. 26, pp. 143–154, 2003.

[45] H. D. Sherali and W. P. Adams, "A hierarchy of relaxations between the continuous and convex hull representations for zero-one programming problems," *SIAM Journal on Discrete Mathematics*, vol. 3, no. 3, pp. 411–430, 1990.

[46] E. Balas, S. Ceria, and G. Cornuéjols, "A lift-and-project cutting plane algorithm for mixed 01 programs," *Mathematical Programming*, vol. 58, pp. 295–324, 1993.

[47] M. X. Goemans and D. P. Williamson, "Improved approximation algorithms for maximum cut and satisfiability problems using semidefinite programming," *Journal of the ACM*, vol. 42, pp. 1115–1145, Nov. 1995.

[48] J. A. Bondy and U. S. R. Murty, *Graph Theory with Applications*. Elsevier Science Ltd, 1976.

[49] D. B. West, *Introduction to Graph Theory*. Prentice Hall, 2000.

[50] E. Lawler, *Combinatorial Optimization: Networks and Matroids*, ser. Dover Books on Mathematics Series. Dover, 1976.

[51] P. Shah and P. Parrilo, "$\mathcal{H}_2$-optimal decentralized control over posets: A state-space solution for state-feedback," *IEEE Transactions on Automatic Control*, vol. 58, no. 12, pp. 3084–3096, Dec. 2013.

[52] R. M. Karp, "Reducibility among combinatorial problems," *Complexity of Computer Computations*, 1972.

[53] G. Rudolph, "Convergence analysis of canonical genetic algorithms," *IEEE Transactions on Neural Networks*, vol. 5, no. 1, pp. 96–101, 1994.

[54] V. Vittal and A. R. Bergen, *Power Systems Analysis*. Prentice Hall, 1999.

[55] H. Saadat, *Power System Analysis*. McGraw-Hill, 2002.

[56] J. Grainger, *Power System Analysis*. McGraw-Hill, 2003.

[57] J. Kirtley, *Electric Power Principles: Sources, Conversion, Distribution and Use*. Wiley, 2011.

[58] A. J. Wood and B. F. Wollenberg, *Power Generation, Operation, and Control*, 3rd ed. Wiley, 2013.

[59] P. Kundur, *Power System Stability and Control*. McGraw-Hill Professional, 1994.

[60] S. Borenstein, M. Jaske, and A. Rosenfeld, "Dynamic pricing, advanced metering, and demand response in electricity markets," UC Berkeley: Center for the Study of Energy Markets, Tech. Rep., 2002.

[61] D. Kirschen, "Demand-side view of electricity markets," *IEEE Transactions on Power Systems*, vol. 18, no. 2, pp. 520–527, 2003.

[62] T. Gedra and P. Varaiya, "Markets and pricing for interruptible electric power," *IEEE Transactions on Power Systems*, vol. 8, no. 1, pp. 122–128, 1993.

[63] K. Fujisawa, M. Kojima, and K. Nakata, "Exploiting sparsity in primal-dual interior-point methods for semidefinite programming," *Mathematical Programming*, vol. 79, pp. 235–253, 1997.

3 Optimal power flow

In the late 1880s, Thomas Edison and George Westinghouse fought the so-called War of the Currents to decide whether the incumbent direct current or Nicola Tesla's alternating current technology would become the standard for future power systems. Endnotes [1, 2] provide excellent historical accounts. The winning argument was that it was much easier using the technology of the time, transformers, to change voltage levels in AC power systems, enabling efficient high-voltage transmission of power and lower voltage generation and end usage (see (2.6) in Section 2.5.1). Beyond pride, part of Edison's opposition to AC transmission was rooted in the higher level of mathematics necessary to understand it.

Today, power electronics have enabled direct current to make a comeback in certain applications like long-distance transmission and microgrids, cf. Section 3.5.1. Some even say that we are now constrained by the mathematical model of AC power flow, which while simple to write down is a quagmire for analysis and computation. Here we tackle this issue head on in one of its purest forms, optimal power flow.

In words, optimal power flow is the problem of minimizing some function of voltage, current, and power, subject to the resulting flow being able to feasibly traverse a transmission or distribution system. Since its introduction by Carpentier [3], virtually every algorithm for continuous optimization has been applied, cf. endnotes [4–12] and the surveys [13–15]. Optimal power flow had similar but separate beginnings in the Russian academic literature [16]. System operators solve optimal power flow routines to do long-term planning, days- to hour-ahead scheduling, real-time dispatch, and pricing (to name a few), making it one of the most frequently employed optimization routines in power systems. As will be seen over the course of this book, many other power system optimizations are essentially optimal power flow models with additional layers of detail.

This chapter will discuss classical linear approximations and modern convex relaxations to the optimal power feasible set. The first instances of relaxations were the SOC formulation of Jabr [17] followed by the SD relaxation of Bai et al. [18]; a large and growing literature has ensued, cf. [19–21]. The reader is referred to endnotes [22, 23] for a comprehensive technical survey. This chapter will also briefly discuss load flow from a somewhat unconventional standpoint and conclude with some minor variations of optimal power flow.

3.1   Basic formulation

Optimal power flow is nearly always formulated in terms of complex voltage, which we denote v_i at node i and v for the column vector. As presented here, optimal power flow could in fact be written exclusively in terms of voltage, but most formulations include additional variables for notational convenience and accessibility. The voltage magnitudes must satisfy the limits

$$\underline{v}_i \leq |v_i| \leq \bar{v}_i.$$

These are usually a small band around a nominal level, such as 0.95 and 1.05 per unit, in place as a conservative stability measure and to ensure that everything connected to the grid sees standard voltage levels.

If node i is connected to node j by a transmission line, the real and reactive power flows between the two are written p_{ij} and q_{ij}; note that these are not the same as the flows arriving at node j from i, p_{ji}, and q_{ji}, which are technically departing node j if positive. Throughout, we will adhere to the convention that departing power flows are positive and arriving power flows negative, i.e., if real power goes from node i to j, $p_{ij} > 0$ and $p_{ji} < 0$. Also, $p_{ij} + p_{ji} \geq 0$ because resistive losses must be nonnegative.

The complex power flow magnitude or *apparent power* must remain below the transmission line's apparent power capacity, $\bar{s}_{ij}$:

$$p_{ij}^2 + q_{ij}^2 \leq \bar{s}_{ij}^2.$$

This is a convex quadratic constraint. Because resistive heating is due to current rather than power, current capacity, sometimes referred to as *ampacity*, is also often used:

$$p_{ij}^2 + q_{ij}^2 \leq |v_i|^2 \bar{I}_{ij}^2. \tag{3.1}$$

Here, $\bar{I}_{ij}$ is the line's maximum allowable current. Example 3.5 will show that this is a convex quadratic constraint in the relaxed coordinate systems of Section 3.3. We will use the apparent power rather then current limit as our default but remark that both are meaningful. In situations where computational tractability is paramount or reactive power is neglected, the linear relaxation

$$|p_{ij}| \leq \bar{s}_{ij}$$

is often used.

The real and reactive powers into or out of node i are the sums of the flows through the transmission lines connected to node i:[1]

$$\sum_j p_{ij} = p_i \quad \text{and} \quad \sum_j q_{ij} = q_i.$$

---

[1] p_i and q_i are usually the difference between the generation and demand, such that $p_i = p_i^g - p_i^d$ and $q_i = q_i^g - q_i^d$ with the generation parts constrained to allowable output ranges and the demand parts constant or similarly constrained. These parameters are absorbed by the generic constraints on p_i and q_i, and so we omit them from our formulations.

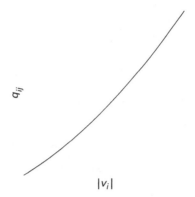

Fig. 3.1 The nonconvex equality constraint relating voltage and power, projected onto voltage magnitude and reactive power variables, $|v_i|$ and q_{ij}.

They are subject to the box constraints

$$\underline{p}_i \le p_i \le \overline{p}_i \quad \text{and} \quad \underline{q}_i \le q_i \le \overline{q}_i.$$

Each line has a complex impedance $z_{ij} = r_{ij} + ix_{ij}$, the reciprocal of which is the admittance, denoted $y_{ij} = g_{ij} - ib_{ij}$. These parameters determine the relationship between power flows and voltages:

$$p_{ij} + iq_{ij} = v_i \left(v_i^* - v_j^* \right) y_{ij}^*,$$

where * denotes the complex conjugate. Cursory background on this model is given in Section 2.5. While simple to write down, this equality constraint is the primary source of nonconvexity in optimal power flow, as illustrated in Figures 2.10 and 3.1.

The final ingredient is the objective function, $f(v, p, q)$, which we assume to be convex but must concede that it often is not. There are multiple relevant choices, including:

- The cost of real power generation, as discussed in Section 2.5.3:

$$\sum_i f_i(p_i).$$

- The total real power generation:

$$\sum_{i \in \mathcal{G}} p_i,$$

where $\mathcal{G}$ is the set of nodes with generators.
- Resistive power losses:

$$\sum_{ij} r_{ij} I_{ij}^2 = \sum_{ij} p_{ij} + p_{ji} = \sum_i p_i.$$

Resistive losses and total real power generation are both linear objectives that often lead to the same optimal solution. However, it is the cost of generation that is most central to power system operations and which is often nonconvex in practice. In this case,

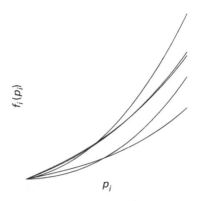

Fig. 3.2   This book operates on the premise that optimal power flow objectives are convex like the above cubic generator supply functions.

the function $f_i(p_i)$ is the system operator's proxy for the cost of producing power at node i. Typically, $f_i(p_i) = 0$ if i is a load, although this may change as loads become more actively involved in power system operations, cf. [24–26]. $f_i(p_i)$ may be an actual physical model of costs, in which case prohibited zones, multiple fuel sources, and valve effects can all contribute to nonconvexity; these issues are addressed in Section 3.5.3. Alternatively, $f_i(p_i)$ may be a supply function submitted by a generation company as a market bid, which can be staircase or other approximations to actual costs [27–29]. We briefly discuss this when we look at electricity markets in Section 6.2. In spite of these realities, the $f_i(p_i)$ are often nearly convex in the sense that their underlying trends resemble convex quadratic or cubic functions like in Figure 3.2, so that any approximation losses are outweighed by computational gains.

Composing these features, optimal power flow is written

$$\underset{v,p,q}{\text{minimize}} \ f(v,p,q) \tag{3.2}$$

$$\text{subject to} \ p_{ij} + iq_{ij} = v_i \left(v_i^* - v_j^* \right) y_{ij}^* \tag{3.3}$$

$$\sum_j p_{ij} = p_i \tag{3.4}$$

$$\sum_j q_{ij} = q_i \tag{3.5}$$

$$\underline{p}_i \leq p_i \leq \overline{p}_i \tag{3.6}$$

$$\underline{q}_i \leq q_i \leq \overline{q}_i \tag{3.7}$$

$$p_{ij}^2 + q_{ij}^2 \leq \overline{s}_{ij}^2 \tag{3.8}$$

$$\underline{v}_i \leq |v_i| \leq \overline{v}_i \tag{3.9}$$

This is a nonconvex QCP. The feasible set associated with this optimization is:

FEASIBLE SET 3.1 (Nonconvex quadratically constrained power flow)   (3.3)-(3.9)

Note that optimal power flow is a steady-state formulation. From a simulation perspective, this description of power system physics is therefore rather minimal in that it omits dynamics, harmonics, and load models, to name a few; moreover, the above is relatively simple even among optimal power flow formulations in that it also omits transformer tap positions and other fine-grained control variables. On the other hand, it is already hovering around the upper tractable limit of optimization, and this is entirely due to (3.3), a (nonconvex) quadratic equality constraint. This alone makes optimizing over Feasible Set 3.1 NP-hard, which here means that there can be no general, efficient algorithm for exactly solving optimal power flow. We will spend the remainder of this chapter developing approximations to (3.3), which we'll then reuse throughout this book.

Example 3.1 *Matrix form of complex power flow.* Constraints (3.3)-(3.5) are commonly expressed as a vector equation using the admittance matrix, defined

$$Y_{ij} = \begin{cases} -y_{ij} & i \neq j \\ \sum_j y_{ij} & i = j \end{cases}.$$

The voltage vector is then related to the real and reactive nodal power vectors by

$$p + iq = \text{diag}[vv^*Y^*],$$

where diag indicates the vector composed of its matrix argument's main diagonal. While substantially more concise, we will generally eschew matrix notation in favor of more intuitive, index-wise exposition.

3.1.1 Nonlinear programming approaches

This book views power systems through convex-colored glasses, so to speak, with the bulk of this chapter being dedicated to convex approximations of optimal power flow. However, Feasible Set 3.1 and variants will always be indispensable because, up to their constitutive modeling assumptions, they represent physical truth. It is therefore essential that we mention NLP approaches to exact, nonconvex optimal power flow. Because this book focuses on modeling and not algorithms, we only very briefly discuss the technical aspects of NLP approaches beyond the statement of Feasible Set 3.1.

Newton's method is at the core of a ubiquitous and powerful group of algorithms and was also among the first applied to optimal power flow in endnote [5]. Many new variants have since appeared, cf. [7, 10, 11, 15, 30], but the basic theory is unchanged; the reader is referred to any of endnotes [31–35] for a detailed discussion on Newton-based interior point methods. Note that SOCPs and SDPs are almost exclusively solved using specialized Newton-based interior point methods, which are distinguished among NLP algorithms because they provably converge in polynomial-time [35, 36]. As discussed above, many other nonlinear programming algorithms have been applied, which we refrain from listing because the Newton-based approaches have been the most successful.

We emphasize that the convex approximations developed in this chapter do not replace Feasible Set 3.1 for exactly the same reason that it has not been replaced by the classical linear approximations of Section 3.2. We would always prefer solutions to Feasible Set 3.1 were they practically obtainable. Unfortunately, optimizing over Feasible Set 3.1 is NP-hard and hence intractable in large instances. When intractability is not an issue, such as in small problem instances or when a starting point near the optimal solution is known, Feasible Set 3.1 should be used. Because such lucky circumstances are rare, the convex approximations in this book are highly practical alternatives. In some special cases like in Section 3.3.1, they produce the same result as Feasible Set 3.1 and thus represent an easier route to the exact solution.

3.2   Linear approximations in voltage-polar coordinates

The polar coordinate representation is the most frequently used due to its intuitive variable definitions and the well-established approximations based on it. In this formulation, voltage is written as the product of a magnitude and a complex exponential, $v = |v|e^{i\theta}$. The resulting feasible set is given by:

FEASIBLE SET 3.2 (Nonconvex voltage-polar coordinate power flow)

$$p_{ij} = g_{ij}|v_i|^2 - |v_i||v_j| \left(g_{ij}\cos(\theta_i - \theta_j) - b_{ij}\sin(\theta_i - \theta_j)\right)$$
$$q_{ij} = b_{ij}|v_i|^2 - |v_i||v_j| \left(g_{ij}\sin(\theta_i - \theta_j) + b_{ij}\cos(\theta_i - \theta_j)\right)$$

$(3.4) - (3.9)$ [Power flow limits]

Both voltage magnitude $|v|$ and angle θ are immediately meaningful quantities, for example, in stability, which we discuss in Section 4.2, and power quality [37, 38]. In particular, it is a minor weakness of the subsequent convex relaxations featured in this book that voltage angles are not explicitly present.

As illustrated in Figures 2.10 and 3.1, the relationship between power and voltage is nonconvex; however, each figure has an approximately linear region, which happens to correspond to nominal operating conditions in which voltage angle differences are small and voltages are nearly one per unit. In this section, we derive approximations based on these observations.

3.2.1   Linearized power flow

The linearized power flow is obtained from the following four approximations:

• All voltage magnitudes are close to one per unit. Set $|v_i| = 1$.
• Conductances are negligible relative to susceptances, i.e., $g_{ij} \ll b_{ij}$. Set $g_{ij} = 0$.

- Voltage angle differences are small enough in magnitude that they occupy the nearly linear region of the sine function, shown in Figure 2.10. Replace $\sin(\theta_i - \theta_j)$ with $\theta_i - \theta_j$. The linearized power flow derives its name from this step because $\theta_i - \theta_j$ is the linear first term in the power series definition of $\sin(\theta_i - \theta_j)$.
- Reactive power flows are negligible relative to real power flows. Remove all reactive power variables and constraints.

The result of these approximations is the following linear feasible set.

FEASIBLE SET 3.3 (Linearized power flow)

$$p_{ij} = b_{ij}(\theta_i - \theta_j)$$

$$\sum_j p_{ij} = p_i$$

$$\underline{p}_i \leq p_i \leq \overline{p}_i$$

$$|p_{ij}| \leq \overline{s}_{ij}$$

If voltage angles are thought of as actual voltages, susceptance as one over resistance, and real power as current, Ohm's law emerges. For this reason, linearized power flow is commonly referred to as "DC power flow." Given the growing number of direct current applications in power systems (see, e.g., Section 3.5.1), we intentionally avoid this terminology in favor of the less ambiguous and literally correct linearized power flow.

Observe that even if we retained the conductances and linearized the cosine in Feasible Set 3.2 by setting it to one, we would still obtain Feasible Set 3.3. However, doing the same in the next section's decoupled power flow would result in bilinear and hence nonconvex constraints, essentially eliminating the purpose of these approximations.

3.2.2 Decoupled power flow

The decoupled power flow of endnote [6] is the above linearized approximation plus a linear approximation of the reactive power flow constraint in Feasible Set 3.2. The first three assumptions of the previous section are applied, except that only one multiplicative factor of $|v_i|$ is set to one in (3.10) rather than both.

FEASIBLE SET 3.4 (Decoupled linear power flow)

$$q_{ij} = b_{ij}\left(|v_i| - |v_j|\right)$$

$$\sum_j q_{ij} = q_i$$

$$\underline{q}_i \leq q_i \leq \overline{q}_i$$

$$\underline{v}_i \leq |v_i| \leq \overline{v}_i$$

Feasible Set 3.3   [Linearized power flow]

The linear and decoupled approximations are indispensable to power system operations: extremely large problems for which either real power is the primary object, such as unit commitment, or for which only ballpark solutions are needed, like in contingency analysis, that often do not admit more detailed approaches under present computational capabilities. Note that the decoupled power flow is somewhat rarely used while the linearized power flow is ubiquitous.

3.2.3   Network flow

We finally consider the simplest possible approximation retaining some notion of flow, *network flow*, which is sometimes also referred to as the transportation model due to its wide usage in the field.

FEASIBLE SET 3.5 (Network flow)

$$p_{ij} + p_{ji} = 0$$
$$q_{ij} + q_{ji} = 0$$

$$(3.4) - (3.7) \quad \text{[Power flow limits]}$$
$$+\text{polyhedral flow capacity constraints}$$

Here we have removed all electrical physics except conservation of power and that flow magnitude cannot exceed line capacity. Note that the flows at either end of a line can be combined into one variable for slightly improved speed. The polyhedral flow capacity constraint is any set of linear constraints that approximates the convex quadratic apparent power constraint (3.8), for example

$$|p_{ij}| + |q_{ij}| \le \sqrt{2}\bar{s}_{ij}$$
$$|p_{ij}| \le \bar{s}_{ij}$$
$$|q_{ij}| \le \bar{s}_{ij}.$$

This is an outer-approximation, which is actually a version of two-commodity network flow. The more common single-commodity network flow, which for our purposes is another real power-only approximation, is obtained by omitting the reactive power constraints (3.5) and (3.7).

While crude for power systems, network flow enjoys extremely fast, specialized algorithms that outperform even linear programming. Moreover, the celebrated *max-flow min-cut theorem* of Ford and Fulkerson [39] provides an exact characterization of network capacity in the single-commodity case. In the present context, it is an often overlooked fact that *single-commodity network flow in radial networks is identical to the linearized power flow because both are fully specified by flow conservation*. The reader is referred to endnotes [40, 41] for comprehensive coverage of network flow.

Example 3.2 *Real power balance.* The simplest possible power flow approximation is the real power balance,

$$\sum_i p_i = 0.$$

This merely requires that the total real power entering the network equals the total power exiting. The real power balance is a relaxation of single-commodity network flow (and hence all other power flow models), which can be seen as follows. Suppose that p_{ij} is a feasible network flow. Then

$$\sum_i p_i = \sum_i \sum_j p_{ij}$$

$$= \frac{1}{2} \sum_i \sum_j p_{ij} + p_{ji}$$

$$= 0.$$

Because any network flow satisfies a real power balance but not every vector that satisfies a real power balance can be realized by a network flow, the real power balance is a relaxation of network flow.

While almost dismissible due to its crudeness, the real power balance is frequently used in practice and is the norm for some extremely hard problems like unit commitment in Section 4.3.

At this point, one might ask: why bother with slower, more accurate methods when the above linear formulations are functional and efficient? An obvious answer is that using more realistic approximations almost always improves real-world performance. The concrete gains are hard to assess in a system as complex as the electric power grid, but a number of qualitative statements can be made:

- Linearizations will always perform poorly away from their operating points, for example during the instabilities that precede many blackouts, which are discussed in Section 4.2.
- The above models neglect losses and couplings between real and reactive power, which are important considerations for planning and operation.
- Perhaps most important, however, is that the next section's relaxations are often not significantly less tractable than the linear approximations and are hence justifiable no matter how small the level of improvement.

3.3   Relaxations

For many years, the linearized approximations of the previous section were the only convex options for power system optimization. Power flow as written in Feasible Set 3.1

is a QCP, which is therefore amenable to the lift-and-project relaxations introduced in Section 2.3.1. By applying the SD relaxation of Example 2.8 followed by the SOC relaxation of Example 2.9, here we obtain SD and SOC relaxations that are often significantly more accurate than the previous linear approximations.

As before, introduce the Hermitian matrix V and the valid constraint $V = vv^*$. This constraint may be equivalently written as the positive semidefiniteness constraint $V \succeq 0$ and rank $V = 1$. If we substitute V_{ij} for each $v_i v_j^*$, we obtain the following equivalent representation of the power flow feasible set:

FEASIBLE SET 3.6 (Nonconvex SD power flow)

$$p_{ij} + iq_{ij} = (V_{ii} - V_{ij})y_{ij}^* \tag{3.10}$$

$$\underline{v}_i^2 \le V_{ii} \le \bar{v}_i^2 \tag{3.11}$$

$$V \succeq 0 \tag{3.12}$$

$$\text{rank } V = 1 \tag{3.13}$$

$(3.4) - (3.8)$ [Power flow limits]

As in Example 2.8, the only nonconvexity is (3.13). We apply the SD relaxation by simply removing it, yielding a convex SD relaxation:

FEASIBLE SET 3.7 (SD complex voltage power flow relaxation)   Feasible Set 3.6 without (3.13).

We will sometimes refer to this relaxation and equivalent feasible sets as the *SD power flow*. If a solution in this feasible set happens to satisfy rank $V = 1$, then it is also in Feasible Set 3.6. If it satisfies rank $V = 1$ and is optimal, then it is optimal for the same objective over Feasible Set 3.6 as well. Because Feasible Set 3.6 is equivalent to the exact, nonconvex optimal power flow (Feasible Set 3.1), such a solution is then a globally optimal power flow solution. Optimal voltages can then be obtained by factoring $V = vv^*$, e.g. via a Cholesky decomposition [42]. Oftentimes an SD relaxation may contain multiple global minima with the same optimal objective value, of which only one has rank one. In such cases, heuristics may be applied to isolate this solution [43].

If rank $V > 1$, the minimal objective over Feasible Set 3.7 is less than or equal to that over Feasible Set 3.6. In this case, the optimal relaxed solution is not in Feasible Set 3.6, and is then an (often very good) approximation of optimal powers and voltage magnitudes. If voltage angles are desired as well, the *principal eigenvector* (which corresponds to the largest eigenvalue) of V is usually a good approximation of the complex voltages. This can be heuristically justified by considering an ideal outcome where rank $V = 1$. In this case, V has only one nonzero eigenvalue, λ, and the eigenvector x must satisfy

$$\lambda x = Vx$$
$$= vv^*x.$$

The only solution in which $\lambda \neq 0$ is $\lambda = v^*v$ and $x = v$. If rank V is not too much larger than one and the principal eigenvector is substantially larger than the others, then we are close to the ideal outcome, and the principal eigenvector serves as a good approximation to the complex voltages.

In applying this relaxation, we have lifted a nonconvex feasible set into a higher dimensional space. Intuitively, we can think of this as "filling in" the nonconvexities; for example, a solid ball is a convex relaxation of a hollow sphere, which is nonconvex. This concept is illustrated for the subsequent branch flow relaxation in Example 3.7. A consequence of this lift is that we now have a complex matrix, V, which has $n(n+1)/2$ distinct entries instead of a complex voltage vector with n distinct entries. In terms of computational complexity, we have essentially squared the number of variables. We also now have a somewhat hefty new constraint, $V \succeq 0$.

Example 3.3 *Matrix form of SD relaxation.* As in Example 3.1, we can consolidate (3.4), (3.5), and (3.10) into the single expression

$$p + iq = \mathrm{diag}\left[VY^*\right].$$

Similar expressions are easily derived for other quantities like losses and objective terms.

Unfortunately, as discussed in Section 2.2.5, SDPs can be challenging to solve at very large scales. One way to make Feasible Set 3.7 more manageable without changing the optimal solution is to recognize that many entries of the matrix V only appear in the semidefinite constraint, particularly in sparser networks. This feature can be exploited by solving the *dual* of the SDP relaxation, which naturally contains the minimum essential number variables [19]. This approach is somewhat inflexible when modeling additional details because duals can be tedious to derive. More significantly, its benefits are lost if the SDP is solved with a *primal-dual* interior point method, which is used by many software packages. Another approach is to directly remove superfluous variables and constraints by exploiting the *sparsity pattern* of the SDP relaxation, which is determined by the underlying network structure of the power system, cf. [44–46].

These approaches may lighten Feasible Set 3.7 but will not change the fact that it must be solved as an SDP. We now construct an SOC relaxation, in which it is trivial to dispose of unnecessary variables and which can be solved using far more efficient SOC interior point methods. We proceed by simply applying the SOC relaxation of Example 2.9 to the SD power flow. Specifically, replace the positive semidefinite constraint (3.12) with the necessary (but not sufficient) condition that each one-by-one and two-by-two principal minor be nonnegative.

FEASIBLE SET 3.8 (SOC complex voltage power flow relaxation) Feasible Set 3.6 without (3.12) and (3.13), and

$$V_{ij} V_{ij}^* \leq V_{ii} V_{jj} \tag{3.14}$$

$$V_{ii} \geq 0 \tag{3.15}$$

We will sometimes refer to this and equivalent feasible sets as the *SOC power flow*. Recall that (3.14) is a special type of SOC constraint called a hyperbolic constraint. Because V_{ii} is real for all i and the left-hand side is always real, (3.14) is a real valued constraint. Note that we have implicitly applied the Hermitian conjugate symmetry constraint $V_{ij} = V_{ji}^*$ by only assigning one variable, V_{ij}, to each node pair.

Observe that if there is no transmission line connecting i and j, (3.14) is the only constraint in Feasible Set 3.8 containing the variable V_{ij}. Hence, we may safely omit (3.14) and V_{ij} for node pairs that do not correspond to lines without affecting the optimal solution. Whereas Feasible Set 3.7 approximately squares the number of variables in the nonconvex QCP formulation, Feasible Set 3.8 adds as many new variables as there are lines, which may be quite modest in realistic systems. Beyond having fewer variables, this relaxation is far more manageable than Feasible Set 3.7 because SOCP is a significantly more tractable optimization class than SDP.

Example 3.4 *Voltage-dependent load modeling in relaxed coordinates.* In Section 2.5.3, we presented the ZIP load model. Although simple, it makes all of our optimal power flow formulations nonconvex. We now cast the ZIP load model in terms of relaxed optimal power flow variables by observing that V_{ii} represents squared voltage magnitude, $|v_i|^2$.

The basic ZIP model can be relaxed by using separate variables for each nodal voltage's magnitude and squared magnitude, $\hat{v}_i$ and V_{ii} respectively. The two variables are then related by the additional constraints

$$\hat{v}_i^2 \leq V_{ii}$$
$$\underline{v}_i \leq \hat{v}_i \leq \bar{v}_i$$
$$p_i = p_i^P + p_i^I \hat{v}_i + p_i^Z V_{ii}$$
$$q_i = q_i^P + q_i^I \hat{v}_i + q_i^Z V_{ii}.$$

It is easy to see that this is a relaxation of the original ZIP model and that the first constraint is a convex quadratic. Hence, they may be inserted into any of the above relaxations without compromising convexity or changing the problem class. If the first constraint is met with equality, the ZIP load model is exact. Unfortunately, this may not occur, in which case $\hat{v}_i$ has no physical meaning.

A second option for loads with no current dependency is simply to use a "ZP" model:

$$p_i = p_i^P + p_i^Z V_{ii}$$
$$q_i = q_i^P + q_i^Z V_{ii}.$$

While less descriptive, this model is fully compatible with the relaxations.

Finally, one can linearize the current portion of the load model by using

$$\hat{v}_i = \sqrt{\overline{V}_i} + \frac{1}{2\sqrt{\overline{V}_i}}\left(V_{ii} - \overline{V}_i\right)$$
$$p_i = p_i^P + p_i^I \hat{v}_i + p_i^Z V_{ii}$$
$$q_i = q_i^P + q_i^I \hat{v}_i + q_i^Z V_{ii},$$

where $\overline{V}_i$ is a constant parameter equal to the nominal value of $|v_i|^2$, e.g. one per unit. We remark that these are very reasonable modeling approximations given the intrinsically approximate nature of ZIP models and, therefore, that they are natural choices for modeling loads in optimal power flow relaxations.

In the above SOC and SD relaxations, we have exchanged nonconvexity for size. In theoretical and practical senses, Feasible Sets 3.7 and 3.8 are easier to work with than exact power flow models because they are convex. In particular, we are guaranteed the following.

- The minimum of any convex objective over Feasible Set 3.7 or 3.8 is a global minimum due to the convexity of SDP and SOCP.
- This minimum objective value is less than or equal to the global minimum of the same objective over Feasible Set 3.6, which is equivalent to exact optimal power flow.
- The SDP and SOCP relaxations can be solved using specialized, polynomial-time interior point methods, which are available in standard software packages.

While these relaxations have performed exceptionally well in empirical studies, they are not always exact, which is to say that systems exist for which there is a nonzero gap between the relaxed and true optimal objectives, cf. [47]. The next section identifies an important class of systems for which they are exact.

3.3.1   Exactness in radial networks

We now examine the theoretical accuracy of the SOC and SD relaxations. Substantial analysis on this topic has shown that, under mild conditions in radial networks, they are exact, which is to say they produce the optimal solution of the nonconvex problem. This section proves a simplified version of this statement; the interested reader is referred to endnotes [20, 22, 23] for broader treatments and to endnote [48] for a more general version of the result for SOC and SD relaxations of QCPs.

Suppose the objective, now just a function of real power, $f(p)$, is convex and strictly increasing. Define the following three optimization problems:

$\mathcal{P}_1$: $\underset{V,p,q}{\text{minimize}} f(p)$ subject to $(V,p,q) \in$ Feasible Set 3.6 (Nonconvex exact)

$\mathcal{P}_2$: $\underset{V,p,q}{\text{minimize}} f(p)$ subject to $(V,p,q) \in$ Feasible Set 3.7 (SD relaxation)

$\mathcal{P}_3$: $\underset{V,p,q}{\text{minimize}} f(p)$ subject to $(V,p,q) \in$ Feasible Set 3.8 (SOC relaxation)

By construction, $\mathcal{P}_1$ is identical to $\mathcal{P}_2$ if the latter's optimal solution happens to satisfy rank $V = 1$. We will show that this is indeed always the case in radial networks and, moreover, that such solutions can be constructed from $\mathcal{P}_3$. Note that here we will assume for notational convenience that $\mathcal{P}_3$ has a full V matrix rather than only the entries corresponding to existing lines.

THEOREM 3.1 *If the network is radial and* $\underline{p}_i = \underline{q}_i = -\bar{s}_{ij} = -\infty$, $\mathcal{P}_2$ *and* $\mathcal{P}_3$ *are exact.*

Proof We will show that any optimal solution of $\mathcal{P}_3$ can be extended to a feasible and hence optimal solution for $\mathcal{P}_1$ and $\mathcal{P}_2$. Suppose that (V,p,q) is optimal for $\mathcal{P}_3$. Observe that if i and j are non-adjacent, i.e., no line exists between them, V_{ij} is a free variable in $\mathcal{P}_3$ except for (3.14).

Assume for the sake of contradiction that (3.14) is not met with equality for some line ij. Then there exists an $\epsilon > 0$ such that $(V_{ij} + \epsilon)(V_{ij}^* + \epsilon) \leq V_{ii}V_{jj}$. Because V_{ij} does not appear in any other SOC constraints, we can substitute $V_{ij} + \epsilon$ for V_{ij} without violating (3.14). We make the following additional substitutions so that (3.4) and (3.10) remain feasible:[2]

$$p_i + iq_i - \epsilon(g_{ij} + ib_{ij}) \longleftrightarrow p_i + iq_i$$
$$p_j + iq_j - \epsilon(g_{ij} + ib_{ij}) \longleftrightarrow p_j + iq_j$$
$$p_{ij} + iq_{ij} - \epsilon(g_{ij} + ib_{ij}) \longleftrightarrow p_{ij} + iq_{ij}$$
$$p_{ji} + iq_{ji} - \epsilon(g_{ij} + ib_{ij}) \longleftrightarrow p_{ji} + iq_{ji}.$$

These substitutions are feasible because $\underline{p}_i = \underline{q}_i = -\bar{s}_{ij} = -\infty$ at each node and line.[3] Because $g_{ij} > 0$ and $f(p)$ is strictly increasing, this reduces the objective value, contradicting the assumption that (V,p,q) is optimal. Therefore, (3.14) must be met with equality for all lines at the optimal solution. Consequently,

$$\det \begin{bmatrix} V_{ii} & V_{ij} \\ V_{ij}^* & V_{jj} \end{bmatrix} = V_{ii}V_{jj} - V_{ij}V_{ij}^*$$
$$= 0$$

for each line. Pairs of voltages corresponding to lines can therefore be extracted using the rank-one decomposition

$$\begin{bmatrix} V_{ii} & V_{ij} \\ V_{ij}^* & V_{jj} \end{bmatrix} = \begin{bmatrix} v_i \\ v_j \end{bmatrix} \begin{bmatrix} v_i \\ v_j \end{bmatrix}^*. \tag{3.16}$$

---

[2] Intuitively, these substitutions remove extraneous line losses, in turn decreasing the power produced by generators and increasing the power received by loads.

[3] In fact, it is easy to show here that the theorem holds for finite limits on real power flows, i.e., $|p_{ij}| \leq \bar{s}_{ij}$: observe that $p_{ij} + p_{ji} > 2\epsilon g_{ij} > 0$ for ϵ to be feasible, which implies $\max\{|p_{ij} - \epsilon g_{ij}|, |p_{ji} - \epsilon g_{ij}|\} < \max\{|p_{ij}|, |p_{ji}|\}$, precluding real power line limit violations.

It remains to show that in radial networks, the collection of these pairs consistently specifies the full vector of voltages, i.e., that if (v_i, v_j) is obtained from one factorization and (v_j, v_k) from another, both produce the same v_j.

Consider an arbitrary path through the network, $\{n_1, \ldots, n_m\}$. Because the network is radial, this is the unique path between nodes n_1 and n_m, and each node on the path is distinct. Choose v_{n_1} such that $|v_{n_1}|^2 = V_{n_1 n_1}$ and $\angle v_{n_1} = 0$. Assume that v_{n_t} is fixed with $|v_{n_t}|^2 = V_{n_t n_t}$ for $n_t = n_1, \ldots, n_k$, $1 \le k \le m-1$. By radiality, $n_{k+1} \notin \{n_1, \ldots, n_k\}$, which guarantees that there is a unique $v_{n_{k+1}}$ for which

$$
\begin{bmatrix} V_{n_k n_k} & V_{n_k n_{k+1}} \\ V^*_{n_k n_{k+1}} & V_{n_{k+1} n_{k+1}} \end{bmatrix} = \begin{bmatrix} v_{n_k} \\ v_{n_{k+1}} \end{bmatrix} \begin{bmatrix} v_{n_k} \\ v_{n_{k+1}} \end{bmatrix}^*
$$

and which does not contradict any prior factorizations; specifically, $v_{n_{k+1}} = V^*_{n_k n_{k+1}} / v^*_{n_k}$. By induction along the path, there exists a set of voltages, $\{v_{n_1}, \ldots, v_{n_m}\}$, each adjacent pair of which satisfies (3.16). Because the path was arbitrary, this holds for any path through the network. Therefore, starting at an arbitrary root node n_r with $|v_{n_r}|^2 = V_{n_r n_r}$ and $\angle v_{n_r}$ fixed at an arbitrary angle, the full voltage vector, v, is obtained by factoring along the unique paths between n_r and each node in the network.[4]

We now specify the free entries of V by also letting $V_{ij} = v_i v_j^*$ for non-adjacent pairs of nodes, which satisfies (3.14) and enables us to write $V = vv^*$. Because V is the outer product of a vector with itself, $V \succeq 0$ and rank $V = 1$. Therefore, V is feasible for $\mathcal{P}_2$ and $\mathcal{P}_1$ in addition to being optimal for $\mathcal{P}_3$. Because $\mathcal{P}_3$ is a relaxation of $\mathcal{P}_2$ and $\mathcal{P}_2$ is a relaxation of $\mathcal{P}_1$, V is optimal for $\mathcal{P}_2$ and $\mathcal{P}_1$. □

Because rank $V = 1$ at an optimal solution for a network in the above-described class, voltages can be obtained from the factorization $V = vv^*$, which will be optimal over Feasible Set 3.1. Because we follow the convention that p_i is negative for loads, the assumption $\underline{p} = \underline{q} = -\infty$ is interpretable as allowing loads to be "over-satisfied" [20, 22, 23].

The essential implication of this theorem is that in radial networks satisfying its assumptions, the SOC relaxation is all that is needed because the nonconvex formulation and the SD relaxation produce the same answers less efficiently. Moreover, because SOCPs can be solved in polynomial-time, optimal power flow can be solved in polynomial-time in radial networks despite its NP-hardness in general networks. More broadly, the theorem is a strong testament to the accuracy of the relaxations: because most real power networks are sparse, it is reasonable to expect the performance degradation to be mild. For comparison, note that the linear approximations in Section 3.2 are never exact for a number of reasons, such as because they neglect losses and reactive power. Indeed, even when the assumptions of Theorem 3.1 are not met, the relaxations are still liable to produce very good approximations.

---

[4] In a non-radial network, loops exist, i.e., paths between nodes and themselves. This sequence of factorizations can fail when a node whose voltage has already been found is reencountered at the end of a loop. In this case, consistent voltage magnitudes but not angles are obtainable.

3.3.2   Real coordinate systems

Feasible Sets 3.7 and 3.8 are complex and thus not immediately processable with any standard software. This is not an issue, however, because any complex SDP or SOCP may be converted to an equivalent real SDP or SOCP. This section shows how to construct a spectrum of real-valued relaxations that includes standard coordinate systems. Mathematically, we will merely formalize a procedure for changing variables.

Define the linear, one-to-one transformation

$$\mathcal{L}(V) = \begin{bmatrix} \text{Re } V & -\text{Im } V \\ \text{Im } V & \text{Re } V \end{bmatrix},$$

and set $W = \alpha^4 Q \mathcal{L}(V) Q^T$, where $\alpha > 0$, $Q \in \mathbb{R}^{2n \times 2n}$, and αQ is orthogonal.[5] Because Im V is skew symmetric, $\mathcal{L}(V)$ and hence W are symmetric.

Pre- and post-multiplying by Q^T and Q, we have

$$\mathcal{L}(V) = Q^T W Q.$$

By splitting $Q^T W Q$ into four submatrices of equal size,

$$Q^T W Q = \begin{bmatrix} A & B^T \\ B & A \end{bmatrix},$$

we can construct the inverse mapping,

$$V = \mathcal{L}^{-1}\left(Q^T W Q\right)$$
$$= A + iB.$$

Observe that $A_{ij} = \left(Q^T W Q\right)_{ij}$ and $B_{ij} = \left(Q^T W Q\right)_{i+n,j}$. The following is a semidefinite relaxation with only real variables:

FEASIBLE SET 3.9 (SD real-valued power flow relaxation)

$$p_{ij} + iq_{ij} = \left(\mathcal{L}^{-1}\left(Q^T W Q\right)_{ii} - \mathcal{L}^{-1}\left(Q^T W Q\right)_{ij}\right) y_{ij}^* \qquad (3.17)$$

$$\left(Q^T W Q\right)_{ij} = \left(Q^T W Q\right)_{i+n,j+n} \qquad (3.18)$$

$$\left(Q^T W Q\right)_{i,j+n} = -\left(Q^T W Q\right)_{i+n,j} \qquad (3.19)$$

$$\underline{v}_i^2 \leq \mathcal{L}^{-1}\left(Q^T W Q\right)_{ii} \leq \overline{v}_i^2 \qquad (3.20)$$

$$W \succeq 0 \qquad (3.21)$$

$$(3.4) - (3.8) \quad \text{[Power flow limits]}$$

(3.18) preserves the equivalence of the top-left and bottom-right blocks, and (3.19) preserves the skew symmetry of the top-right and bottom-left blocks. When split into real and imaginary parts, (3.17) is written

---

[5] A real matrix S is orthogonal if $S^T S = SS^T = I$. Here, $\alpha^2 Q^T Q = \alpha^2 Q Q^T = I$.

$$p_{ij} = g_{ij} \left(Q^T W Q \right)_{ii} - g_{ij} \left(Q^T W Q \right)_{ij} + b_{ij} \left(Q^T W Q \right)_{i+n,j}$$
$$q_{ij} = b_{ij} \left(Q^T W Q \right)_{ii} - g_{ij} \left(Q^T W Q \right)_{i+n,j} - b_{ij} \left(Q^T W Q \right)_{ij},$$

where $\left(Q^T W Q \right)_{i+n,i} = 0$ due to the symmetry of $Q^T W Q$ and (3.19).

The following lemma shows that the complex- and real-valued SD relaxations are equivalent.

LEMMA 3.1 *Let* $W = \alpha^4 Q \mathcal{L}(V) Q^T$, *where* αQ *is orthogonal and* $\alpha > 0$. *V is in Feasible Set 3.7 if and only if W is in Feasible Set 3.9.*

Proof We must show that $W \succeq 0$ if and only if $V \succeq 0$. From endnote [49], $\mathcal{L}(V) \succeq 0$ if and only if $V \succeq 0$. Because eigenvalue sign and hence positive semidefiniteness are invariant under positively scaled orthogonal transformations, we also have that $\alpha^4 Q \mathcal{L}(V) Q^T \succeq 0$ if and only if $V \succeq 0$, which establishes the claim. □

Just as for the complex SD relaxation, a solution that minimizes an objective over Feasible Set 3.9 corresponds to the global minimum of that objective over the nonconvex, exact feasible set if rank $W = 1$. In this case, voltages can be recovered by factoring $\mathcal{L}^{-1} \left(Q^T W Q \right) = vv^*$.

The corresponding real-valued SOC relaxation is obtained by replacing $W \succeq 0$ with

$$\mathcal{L}^{-1} \left(Q^T W Q \right)_{ij} \mathcal{L}^{-1} \left(Q^T W Q \right)_{ij}^* \leq \mathcal{L}^{-1} \left(Q^T W Q \right)_{ii} \mathcal{L}^{-1} \left(Q^T W Q \right)_{jj} \tag{3.22}$$

$$\mathcal{L}^{-1} \left(Q^T W Q \right)_{ii} \geq 0, \tag{3.23}$$

which yields the SOC relaxation:

FEASIBLE SET 3.10 (SOC real-valued power flow relaxation) Feasible Set 3.9 with (3.22) and (3.23) instead of (3.21).

It can be shown that Feasible Sets 3.8 and 3.10 are similarly related via an argument nearly identical to that for Lemma 3.1. Observe that the same SOC relaxation technique could instead be directly applied to $W \succeq 0$ in Feasible Set 3.9, yielding an equivalent SOC relaxation.

Example 3.5 *Current limits in relaxed coordinates.* In any relaxed coordinate system, each node's squared voltage magnitude is a linear function of the relaxation variables. For instance, it is simply V_{ii} in the complex SOC and SD relaxation, and it is $\mathcal{L}^{-1} \left(Q^T W Q \right)_{ii}$ in a real-valued coordinate system. This enables us to write the transmission current capacity or ampacity limit, (3.1), as

$$p_{ij}^2 + q_{ij}^2 \leq \mathcal{L}^{-1} \left(Q^T W Q \right)_{ii} \bar{I}_{ij}^2.$$

Because this is a convex quadratic constraint, it can be seamlessly incorporated into an SOC or SD relaxation without changing the optimization class.

This procedure changes the variables of the complex power flow relaxations into real-valued ones that can be processed by standard software. If the purpose is merely to solve optimal power flow, the choice of the orthogonal matrix Q is not critical because there is a one-to-one mapping between any two coordinate systems. However, some coordinate systems may be better suited than others for certain applications, or for deriving further approximations. We now present the relaxation in two standard coordinate systems.

Voltage-polar coordinates

The simplest choice of Q is the identity matrix,

$$Q = \begin{bmatrix} I & 0 \\ 0 & I \end{bmatrix},$$

which is trivially orthogonal. In this case, (3.22) is written

$$W_{ij}^2 + W_{i+n,j}^2 \le W_{ii} W_{jj}, \tag{3.24}$$

which is exactly the original SOC relaxation of [17]. The full feasible set for the SD relaxation in this coordinate system is written below.

FEASIBLE SET 3.11 (SD real power flow relaxation in voltage-polar coordinates)

$$p_{ij} = g_{ij} W_{ii} - g_{ij} W_{ij} + b_{ij} W_{i+n,j}$$
$$q_{ij} = b_{ij} W_{ii} - g_{ij} W_{i+n,j} - b_{ij} W_{ij}$$
$$W_{ij} = W_{i+n,j+n}$$
$$W_{i,j+n} = -W_{i+n,j}$$
$$\underline{v}_i^2 \le W_{ii} \le \bar{v}_i^2$$
$$W \succeq 0$$

$$(3.4) - (3.8) \quad \text{[Power flow limits]}$$

Just as entries of the Hermitian matrix V were substituted for products of voltages in the complex relaxations, real-valued relaxations can also be directly constructed from real-valued power flow formulations like Feasible Set 3.2. The above relaxations are obtained from the substitutions

$$W_{ij} \longleftrightarrow |v_i||v_j| \cos(\theta_i - \theta_j)$$
$$W_{i+n,j} \longleftrightarrow |v_i||v_j| \sin(\theta_i - \theta_j).$$

Due to this correspondence, we refer to this as the "voltage-polar coordinate" relaxation.

Example 3.6 *Resistive losses in the voltage-polar coordinate relaxation.* All variables in this section's convex relaxations are commensurate with squared voltages. Starting from first principles, power quantities such as resistive losses thus simplify to linear expressions, as given below:

$$\sum_{ij} g_{ij} \left(v_i - v_j \right) \left(v_i - v_j \right)^* = \sum_{ij} g_{ij} \left(|v_i|^2 + |v_j|^2 - 2|v_i||v_j| \cos(\theta_i - \theta_j) \right)$$

$$\longleftrightarrow \sum_{ij} g_{ij} \left(W_{ii} + W_{jj} - 2W_{ij} \right).$$

Of course, there is no benefit in using this over $\sum_{ij} p_{ij} + p_{ji}$ because both are linear.

Voltage-rectangular coordinates

Now consider the matrix

$$Q = \begin{bmatrix} I & I \\ -I & I \end{bmatrix},$$

and note that $Q/\sqrt{2}$ is orthogonal. If we write voltage as $v = w + ix$, this "voltage-rectangular coordinate" relaxation corresponds to the substitutions

$$W_{ij} \longleftrightarrow w_i w_j$$
$$W_{i+n,j} \longleftrightarrow x_i w_j$$
$$W_{i,j+n} \longleftrightarrow w_i x_j$$
$$W_{i+n,j+n} \longleftrightarrow x_i x_j,$$

which matches the original SD relaxation of endnote [18]. The feasible set is given below.

FEASIBLE SET 3.12 (SD real power flow relaxation in voltage-rectangular coordinates)

$$p_{ij} = g_{ij}(W_{ii} + W_{i+n,i+n}) - g_{ij}(W_{ij} + W_{i+n,j+n})$$
$$+b_{ij}(W_{j,i+n} - W_{i,j+n}) \tag{3.25}$$
$$q_{ij} = b_{ij}(W_{ii} + W_{i+n,i+n}) - g_{ij}(W_{j,i+n} - W_{i,j+n})$$
$$-b_{ij}(W_{ij} + W_{i+n,j+n}) \tag{3.26}$$
$$W_{i,j+n} = -W_{i+n,j} \tag{3.27}$$
$$W_{ij} = W_{i+n,j+n} \tag{3.28}$$
$$\underline{v}_i^2 \le W_{ii} + W_{i+n,i+n} \le \overline{v}_i^2 \tag{3.29}$$
$$W \succeq 0 \tag{3.30}$$

$(3.4) - (3.8)$ [Power flow limits]

Coincidentally, the structural constraints (3.27) and (3.28) come out the same as in the voltage-polar case. If rank $W = 1$, voltages can be retrieved either using the factorization discussed above or equivalently from the factorization

$$W = \begin{bmatrix} w \\ x \end{bmatrix} \begin{bmatrix} w \\ x \end{bmatrix}^T .$$

In the latter case, the voltage vector is simply $v = w + ix$.

3.3.3 Branch flow models

The branch flow model was initially presented under the name *DistFlow equations* for reconfiguration [50] and capacitor placement [51] in radial distribution systems, topics covered in Sections 4.4 and 5.1.2, respectively. For approximately twenty years following their introduction, they remained an obscure radial power flow formulation. Indeed, when accustomed to voltage-centric formulations, the resulting equations appear somewhat peculiar in that the variables are the line flows p_{ij} and q_{ij}, squared voltage magnitudes $|v_i|^2$, and squared current magnitudes, the latter two of which we write v_i and ψ_{ij}. Convexity was identified in endnotes [52–54], and the corresponding SOC relaxation has since been shown to be equivalent to the preceding "nodal" relaxations [22]. Our interest in the branch flow model is therefore not to obtain yet another relaxation but for the alternate insights and subsequent approximations it admits.

Because the branch flow equations are not commonly seen in the literature or immediately obvious, we derive them now. Consider the complex power loss from node i to node j,

$$|I_{ij}|^2 z_{ij} = \frac{p_{ij}^2 + q_{ij}^2}{|v_i|^2} z_{ij},$$

which we rewrite in terms of our new variables as

$$\psi_{ij} z_{ij} = \frac{p_{ij}^2 + q_{ij}^2}{v_i} z_{ij}.$$

Because we are following the convention that departing power is positive and arriving negative, the real and reactive line losses are $p_{ij} + p_{ji}$ and $q_{ij} + q_{ji}$, which are always positive (assuming $r_{ij} > 0$ and $x_{ij} > 0$). We may then write

$$\psi_{ij} v_i = p_{ij}^2 + q_{ij}^2 \tag{3.31}$$

$$p_{ij} + p_{ji} = r_{ij} \psi_{ij} \tag{3.32}$$

$$q_{ij} + q_{ji} = x_{ij} \psi_{ij}. \tag{3.33}$$

Here, ψ_{ij} is the squared magnitude of the current on line ij, which is proportional to the real and reactive power losses. Knowing the power out of and voltage at node i, the above equations define the power through line ij into node j. The voltage at node j is given by Ohm's law:

$$v_j = v_i - I_{ij} z_{ij}. \tag{3.34}$$

Equations (3.31)-(3.34) with nodal power conservation and line flow limits constitute an exact description of power flow, equivalent to Feasible Set 3.1.

To convexify, we begin by taking the squared complex magnitude of each side of (3.34):

$$v_j = |v_i - I_{ij}z_{ij}|^2$$
$$= v_i - 2\text{Re}\left[v_i^* I_{ij}z_{ij}\right] + |I_{ij}|^2 |z_{ij}|^2$$
$$= v_i - 2\left(r_{ij}p_{ij} + x_{ij}q_{ij}\right) + \left(r_{ij}^2 + x_{ij}^2\right)\psi_{ij}.$$

At this point, the only remaining variable quantities are the line flows p_{ij} and q_{ij}, squared voltage magnitude v_i, and squared current magnitude ψ_{ij}.

Now the only remaining nonconvexity is (3.31). Increased current and hence line losses should not improve most realistic objectives, and therefore changing the equality in (3.31) to "$\geq$" should not change the resulting optimum. We thus relax (3.31) to

$$p_{ij}^2 + q_{ij}^2 \leq \psi_{ij}v_i. \tag{3.35}$$

But this is just a hyperbolic constraint, which we showed in Example 2.9 to be a garden variety SOC constraint. In the present case, (3.35) is written in standard SOC form as

$$\left\| \begin{bmatrix} 2 & 0 & 0 & 0 \\ 0 & 2 & 0 & 0 \\ 0 & 0 & 1 & -1 \end{bmatrix} \begin{bmatrix} p_{ij} \\ q_{ij} \\ \psi_{ij} \\ v_i \end{bmatrix} \right\| \leq [0 \ \ 0 \ \ 1 \ \ 1] \begin{bmatrix} p_{ij} \\ q_{ij} \\ \psi_{ij} \\ v_i \end{bmatrix}.$$

We have now made all constraints convex and collect them below for convenience. The reader is reminded that v_i and ψ_{ij} represent the squared voltage and current magnitudes, $|v_i|^2$ and $|I_{ij}|^2$.

FEASIBLE SET 3.13 (SOC branch flow relaxation)

$$p_{ij}^2 + q_{ij}^2 \leq \psi_{ij}v_i \tag{3.36}$$
$$p_{ij} + p_{ji} = r_{ij}\psi_{ij} \tag{3.37}$$
$$q_{ij} + q_{ji} = x_{ij}\psi_{ij} \tag{3.38}$$
$$v_j = v_i - 2\left(r_{ij}p_{ij} + x_{ij}q_{ij}\right) + \left(r_{ij}^2 + x_{ij}^2\right)\psi_{ij} \tag{3.39}$$
$$\underline{v}_i^2 \leq v_i \leq \overline{v}_i^2 \tag{3.40}$$

$$(3.4) - (3.8) \quad \text{[Power flow limits]}$$

We have derived a convex relaxation starting from the branch flow equations, but what exactly has been relaxed, since we expect that the SOC constraint will be met with equality? As in the relaxations of Section 3.3, voltage angles are lost when taking the squared magnitude of each side of (3.34), effectively enforcing only the magnitude portion of the original constraint. A simple instance of the relaxation is examined in Example 3.7.

Example 3.7 *Visualization of the branch flow relaxation with an infinite bus.* The infinite bus is a common approximation for modeling a single component like a generator connected by a transmission line to a significantly larger power system. The approximation is applied by modeling the larger system as a single node whose voltage magnitude and frequency, $|v_b|$ and $\dot{\theta}_b$, are insensitive to the connected component (frequency is not relevant to this example).

In this example, we use an infinite bus approximation to visualize the branch flow relaxation. Suppose that a generator at node i is sending constant real power, $p_{ib} = 1$, through a transmission line with resistance $r_{ib} = 0.01$ to an infinite bus with $|v_b| = 1$, and hence $v_b = 1$. The combined exact and relaxed real power constraints on the real and reactive power received by the infinite bus take the form

$$r_{ib}\left(p_{bi}^2 + q_{bi}^2\right) = 1 + p_{bi} \quad \text{and} \quad r_{ib}\left(p_{bi}^2 + q_{bi}^2\right) \leq 1 + p_{bi},$$

respectively. The exact constraint is a nonconvex quadratic equality, which is a one-dimensional curve on the (p_{bi}, q_{bi}) plane. In Figure 3.3, the relaxation opens up the nonconvex constraint into a two-dimensional subset of the plane, which is convex.

Quadratic approximation to the branch flow model

Recall that in Section 3.2.1, one of the assumptions leading to the linearized power flow was $|v_i| = 1$ for all i. The same assumption may be applied in isolation to the branch flow model by setting $v_i = 1$, yielding a very concise CQCP approximation from endnote [55]:

FEASIBLE SET 3.14 (CQCP branch flow approximation)

$$r_{ij}\left(p_{ij}^2 + q_{ij}^2\right) \leq p_{ij} + p_{ji}$$

$$x_{ij}\left(p_{ij}^2 + q_{ij}^2\right) \leq q_{ij} + q_{ji}$$

$$(3.4) - (3.8) \quad \text{[Power flow limits]}$$

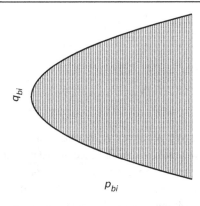

Fig. 3.3   The solid line is the set of feasible points described by the nonconvex exact constraint, and the enclosed shaded area is the set of points that satisfy the convex relaxed constraint for $-0.99 \leq p_{bi} \leq -.9$.

Note that we have combined (3.36) with (3.37) and (3.38), eliminating the need for the variables ψ_{ij}. This approximation increases the objective by respectively fixing voltage magnitudes at one and subsequently decreases it by removing the voltage equation. Because this is not a relaxation but an approximation, it can attain objectives that are higher or lower than the true minimum. While less accurate than the above convex relaxations, it is still substantially more accurate than the linearized approximation because losses and reactive power are accounted for.

As stated in Section 2.2.2, SOCP algorithms are used to solve CQCPs, so there is little computational return on this approximation at face value other than that from reducing the number of variables. However, the simplification makes room for other modeling details, which we'll incorporate later when we do transmission system planning in Section 5.2.

Finally, observe that quadratic approximations can be obtained by setting voltage magnitudes equal to one in nodal SOC relaxations. For instance, replacing (3.14) in Feasible Set 3.8 with

$$V_{ij} V_{ij}^* \leq 1$$

leads to an equivalent quadratic relaxation. Whether to do so in a nodal or branch flow model is merely a matter of preference.

3.3.4 Further discussion

The relaxations introduced in this section are the basic building blocks of subsequent chapters. We now momentarily look back to discuss some of their salient characteristics and practical implications. First, we assemble the relaxations and the techniques leading to them in Figure 3.4. Note that the branch flow relaxation sits with the other real-valued SOC relaxations in the bottom-right box.

If rank $V = 1$ in the SD relaxation, then complex voltages are obtainable from the factorization $V = vv^*$, for example via Cholesky decomposition. If this condition is not satisfied, it may not be possible to extract consistent voltage angles from the relaxations. This limitation provides some intuition beyond the removal of the rank constraint: the relaxations drop the physical requirement that voltage angles around a loop must sum to zero. This is somewhat formalized by Theorem 3.1, which essentially says that the relaxations lose no information in networks with no loops.

A relaxed power flow solution is a useful approximation even when it is infeasible for the exact, nonconvex problem. Indeed, all real and reactive powers and voltage magnitudes are specified in each relaxation, and complex voltages are approximated by the principal eigenvector of V. As we will see over the course of this book, knowing these quantities approximately is sufficient for many applications. The relaxations also contain far more information than the linearized approximations of Section 3.2, which virtually never produce feasible solutions and yet are widely accepted. Lastly, relaxations are innately more informative than general approximations because they provide bounds, which we elaborate on below.

Recall that in Section 2.3 we observed the useful consequence that a feasible solution and a relaxation together provide a two-sided bound on the optimal objective of an

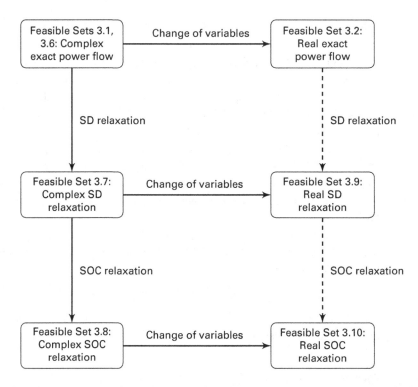

Fig. 3.4   Relationships among power flow relaxations. Those indicated by dotted lines were not explicitly covered.

optimization problem. We now restate this fact in terms of optimal power flow. Suppose (v', p', q') is in Feasible Set 3.1, the exact, nonconvex power flow. Then we have the following inequalities:

$$\min_{V,p,q} f(V,p,q) \quad \text{subject to} \quad (V,p,q) \in \text{Feasible Set 3.8}$$

$$\leq \min_{V,p,q} f(V,p,q) \quad \text{subject to} \quad (V,p,q) \in \text{Feasible Set 3.7}$$

$$\leq \min_{V,p,q} f(V,p,q) \quad \text{subject to} \quad (V,p,q) \in \text{Feasible Set 3.6}$$

$$= \min_{v,p,q} f(v,p,q) \quad \text{subject to} \quad (v,p,q) \in \text{Feasible Set 3.1}$$

$$\leq f(v',p',q')$$

The inequality is illustrated in Figure 3.5. This inequality is of great practical use: because the relaxed solutions and (v', p', q') will generally be considerably easier to obtain than the true optimum, they provide a tractable means of approximating nonconvex optimal power flow and assessing the quality of the approximation. If $f(v', p', q') = f(V, p, q)$, the global optimality of the feasible point (v', p', q') is certified, which is in general hard to do using only the nonconvex formulation. Note that bounds such as these cannot be inferred from the linearized power flow approximation of Section 3.2.1, which may lead to higher or lower objectives than the true optimum.

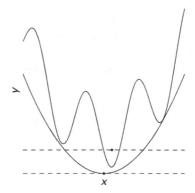

Fig. 3.5   An illustration of the bound implied by a feasible solution and relaxed minimum on a true minimum. Here, the curves represent a nonconvex feasible set and its convex relaxation, and the objective function is $f(x, y) = y$.

3.4   Load flow

Even more fundamental than optimal power flow is the problem of solving for voltages and power flows given specified nodal characteristics. This is referred to simply as *power flow* or *load flow* [27]; to minimize the risk of ambiguity, we only use the latter terminology. Although load flow is not an optimization problem at face value, it is heavily relied on by system operators and is amenable to many of the same tools already employed in this chapter.

In standard load flow, there are three types of nodes:

- Nodes where p and $|v|$ are specified. These are usually power producers.
- Nodes where p and q are specified. These are usually loads.
- The *swing bus*, at which $v_s = 1$.

There may be an infinite number of load flow solutions because identically translating all voltage angles does not change the other variables. The swing bus is a nonphysical convention used to select the distinct solution for which an arbitrary node's complex voltage angle is zero and magnitude is one. Note that in linearized optimal power flow, this is not a useful convention because it has no effect on the resulting power flows. In more accurate relaxed or exact optimal power flow formulations, it is neither valid nor useful because otherwise good solutions may become artificially infeasible.

3.4.1   Exact load flow

Finding exact power flows entails solving (3.3)-(3.5) given the above-described producer and load node parameters. Load flow has traditionally been solved entirely numerically using techniques like the Gauss-Seidel and Newton-Raphson methods. These approaches have been honed over many years and are now reasonably robust and scalable. Here, we instead present an alternative formulation in the spirit of the relaxations of Section 3.3. Note that the following does not represent the standard or even most effective approach to load flow, but it provides an important conceptual link to the rest of this chapter.

Denote the swing bus by s and the set of generators and loads by $\mathcal{G}$ and $\mathcal{L}$, respectively. It is most convenient to our formulation to use voltage-rectangular coordinates because the swing bus s can be specified via $w_s = 1$ and $x_s = 0$, which is enforced through (3.43) and (3.44). Let hatted quantities denote those fixed as discussed in the beginning of this section. The exact load flow problem is then as follows.

$$\underset{W,p,q}{\text{solve}} \quad p_i = \hat{p}_i, \quad W_{ii} = |\hat{v}_i|^2 \quad \text{for } i \in \mathcal{G} \tag{3.41}$$

$$p_i = \hat{p}_i, \quad q_i = \hat{q}_i \quad \text{for } i \in \mathcal{L} \tag{3.42}$$

$$W_{ss} = 1 \tag{3.43}$$

$$W_{s+n,i} = W_{i,s+n} = W_{s+n,s+n} = 0 \tag{3.44}$$

$$W \succeq 0 \tag{3.45}$$

$$\text{rank } W = 1 \tag{3.46}$$

$(3.4), (3.5) \quad$ [Nodal power balance]

$(3.25) - (3.28) \quad$ [Voltage-rectangular coordinate SD power flow]

Exactly as in Section 3.3, the SD relaxation is applied by removing the rank constraint. However, we now have a feasible set with a continuum of feasible solutions but no objective. The most obvious choice is the objective rank W, because any solution that has the lowest possible rank, one, is a feasible power flow. Thus, if a feasible power flow exists, it must be the optimal solution, and so this is an equivalent formulation of load flow.

Unfortunately, rank minimization is NP-hard. Minimizing the trace of W is good approximation to rank minimization, yielding the below SDP:

$$\underset{W,p,q}{\text{minimize}} \text{ tr } W$$
$$\text{subject to } (3.4), (3.5), (3.25) - (3.28), (3.41) - (3.45)$$

But this is essentially optimal power flow with a linear, increasing objective and thus inherits Theorem 3.1. As such, in a radial network, we could replace the SD constraint (3.45) with its SOC relaxation. If rank $W = 1$, voltages in the form $v = w + ix$ are obtained from the factorization

$$W = \begin{bmatrix} w \\ x \end{bmatrix} \begin{bmatrix} w \\ x \end{bmatrix}^T.$$

This load flow approximation has some interesting implications. Foremost, it is an approximation and thus is not guaranteed to produce the exact load flow solution, which could be problematic in applications demanding precision. However, the non-approximate formulation (with the rank objective) is NP-hard, implying that load flow is fundamentally hard and bringing into question the theoretical efficacy of traditional approaches like the Gauss-Seidel and Newton-Raphson methods.

This load flow approximation also happens to be a special case of a more general procedure for solving quadratic equations known as *PhaseLift* [56], which has been shown to have a number of nice theoretical properties. PhaseLift is more similar to traditional methods than is immediately apparent, because SDPs are usually solved using interior point methods that are based on Newton's method, a relative of the Newton-Raphson method [33].

There is considerable room for future work on this topic. Indeed, basic issues such as existence and uniqueness of load flow solutions remain at large [57, 58]; notably, it is shown in Chiang and Baran [59] that a unique load flow always exists in radial networks, which is analogous to Theorem 3.1. It is likely that the PhaseLift perspective may provide more analytical insight in more general scenarios than traditional approaches.

3.4.2   Linearized load flow

Within the linearized power flow model of Section 3.2.1, load flow simplifies to the solution of a rank deficient linear system. Define the inductance matrix

$$B_{ij} = \begin{cases} -b_{ij} & \text{if } i \neq j \\ \sum_j b_{ij} & \text{if } i = j \end{cases}.$$

As an aside, note that B may be viewed as the Laplacian matrix of a graph with edge weights equal to line inductances. A rich theory surrounds the eigenvalues of Laplacian matrices of graphs, collectively known as *spectral graph theory* [60].

It is straightforward to show that B (and all graph Laplacians) has rank $n - 1$. The linear relationship between nodal power and voltage angle is expressed by the matrix equation

$$p = B\theta.$$

Note that this is just the two equality constraints in Feasible Set 3.3 written in matrix form.

The goal is to find the vector θ given p, from which the line flows are calculated as

$$p_{ij} = b_{ij}(\theta_i - \theta_j).$$

Because the linearized power flow is lossless, p must satisfy power conservation,

$$\sum_i p_i = 0.$$

Let λ_i and γ_i denote the ith eigenvalue and eigenvector of B, and define the *Moore-Penrose pseudoinverse* [61] of B to be

$$B^\dagger = \sum_{i:\lambda_i \neq 0} \frac{1}{\lambda_i} \gamma_i \gamma_i^T.$$

A feasible set of voltage angles is given by

$$\theta = B^\dagger p.$$

This might be said to be the "minimum energy solution" because it solves the minimization

$$\underset{\theta}{\text{minimize}} \sum_i \theta_i^2 \quad \text{subject to} \quad p = B\theta.$$

The swing bus convention is applied by choosing an individual node s and subtracting θ_s from each element of θ; note that θ and $\theta - \theta_s$ are equally valid. Furthermore, because the vector of all ones is the null space of B, any uniform translation of θ is feasible.

Another perfectly valid approach is to first designate $\theta_s = 0$, invert the full rank submatrix of B obtained by removing the sth row and column, and then multiply the corresponding subvector of p by the inverted submatrix. This approach directly produces the solution with $\theta_s = 0$.

3.5   Extensions

This section discusses several light additions and alterations to basic optimal power flow.

3.5.1   Direct current networks

There has been a resurgence of interest in direct current power infrastructures, ranging from low-voltage industrial and commercial settings [62, 63] and semi-autonomous *microgrids* [64, 65] to *high-voltage direct current* transmission lines [66]. Most such applications are enabled by modern power electronic technologies [67], which were not available until 100 years after the War of the Currents was settled in favor AC power. This begs the question: If we could start over from scratch, would an AC or DC system be better [68]? It's a complex discussion outside the scope of this book, except for the obvious point that we won't discard an infrastructure that costs trillions of dollars just to find out. While none of the above applications yet qualifies as a power system on which one would solve optimal power flow problems, it is a plausible future scenario, and subsequent topics in this book do already have relevance in DC settings.

The nonconvex DC power flow feasible set is as follows.

FEASIBLE SET 3.15 (Nonconvex quadratically constrained direct current power flow)

$$p_{ij} = v_i(v_i - v_j)/r_{ij} \tag{3.47}$$

$$\sum_j p_{ij} = p_i \tag{3.48}$$

$$\underline{p}_i \leq p_i \leq \overline{p}_i \tag{3.49}$$

$$|p_{ij}| \leq \overline{s}_{ij} \tag{3.50}$$

$$\underline{v}_i \leq v_i \leq \overline{v}_i \tag{3.51}$$

It is actually a misnomer to present direct current power flow as an extension because the above model is in fact a restriction of the earlier AC formulations. Just set $\underline{q}_i = \overline{q}_i = b_{ij} = 0$ for all i and j and Feasible Set 3.15 drops out of Feasible Set 3.1.[6]

Because all quantities are real, there is no need to consider multiple relaxed coordinate systems as in Section 3.3.2. Proceeding as in Section 3.3, we apply the SD relaxation to obtain the relaxed feasible set.

FEASIBLE SET 3.16 (SD direct current power flow relaxation) (3.48)-(3.50) and

$$p_{ij} = (V_{ii} - V_{ij})/r_{ij} \tag{3.52}$$
$$\underline{v}_i^2 \leq V_{ii} \leq \overline{v}_i^2 \tag{3.53}$$
$$V \succeq 0 \tag{3.54}$$

Unlike the AC case, the above relaxation is always tight if $\underline{p} = -\infty$ and the objective is increasing, regardless of network topology [19]. We obtain an SOC relaxation by applying the SOC relaxation as before.

FEASIBLE SET 3.17 (SOC direct current power flow relaxation) (3.48)-(3.50), (3.52)-(3.53), and

$$V_{ij}^2 \leq V_{ii} V_{jj} \tag{3.55}$$

If $\overline{v} = \infty$ and the objective is increasing, both the SD and SOC relaxations are exact [69].

3.5.2   Reactive power capability curves

While the main purpose of generators is real power production, they are also primary contributors to power system stability, an important part of which is reactive power. In some senses, the cost of reactive power is zero because it carries no energy. However, reactive power occupies transmission line and generator capacities, limiting the amount of real power a generator can output. This causes generators providing reactive power to incur *opportunity costs* [70, 71], which are the forfeited profits from real power they cannot sell.

These opportunity costs can be straightforwardly incorporated into optimal power flow routines via additional constraints comprising the reactive power capability curve [72]. They are concisely expressed within convex formulations as a combination of one linear and two convex quadratic constraints. Let x_i, $\overline{I}_i$, $\overline{E}_i$, and $\overline{\theta}_i$ be generator i's synchronous reactance and maximum allowable armature current, internal voltage,

---

[6] While mathematically correct, note that this perspective is conceptually inconsistent in that the resulting variables are no longer phasors representing sinusoidal functions but real numbers representing constant functions.

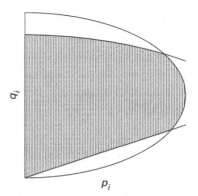

Fig. 3.6   The convex set described by a reactive power capability curve, projected onto real and reactive power variables, p_i and q_i.

and angle between terminal voltage and the quadrature axis, respectively. The three capability curve constraints are:

- Underexcitation limit:

$$\frac{p_i}{\tan \overline{\theta}_i} \leq q_i + \frac{|v_i|^2}{x_i}$$

- Maximum armature current:

$$p_i^2 + q_i^2 \leq |v_i|^2 \overline{I}_i^2$$

- Maximum field heating:

$$\left(|v_i|^2 + x_i q_i \right)^2 + x_i^2 p_i^2 \leq |v_i|^2 \overline{E}_i^2$$

The region prescribed by the constraints is illustrated in Figure 3.6. Now recall that the squared voltage magnitude, $|v_i|^2$, is a linear function of the variables in all of the relaxations we've introduced, for example $W_{ii} + W_{i+n,i+n}$ in the voltage-rectangular coordinate SD relaxation (Feasible Set 3.12) and v_i in the SOC branch flow relaxation (Feasible Set 3.13). Hence, the above three curve constraints are respectively linear, convex quadratic, and convex quadratic in any relaxed coordinate system. They may be embedded in a relaxed power flow feasible set without compromising convexity and at virtually no additionally computational cost. This is quite useful for pricing reactive power, which we discuss later in Example 6.6.

3.5.3   Nonconvex generator cost curves

Two first-order sources of nonconvexity in generator cost curves are fuel source transitions between different real power output levels and valve-point loading effects resulting from multiple steam valves opening sequentially. The former, shown in Figure 3.7, is typically modeled with piecewise quadratic functions [73, 74], and the latter via the addition of the absolute value of a sinusoid [75]. We may straightforwardly address

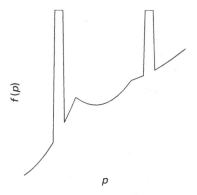

$f(p)$

p

Fig. 3.7   A fuel curve with nonconvexities due to multiple fuel sources and prohibited zones of operation.

the former within the framework of convex optimization, and the latter is hopelessly nonconvex. In doing so, we make our first use of integer variables as a modeling tool.

The composite cost function of generator i switching between fuel sources $1, \ldots, n_i$ may be expressed

$$f_i(p_i) = \begin{cases} a_i^1 p_i^2 + b_i^1 p_i + c_i^1, & \underline{p}_i^1 \le p_i \le \overline{p}_i^1 \\ \vdots & \vdots \\ a_i^{n_i} p_i^2 + b_i^{n_i} p_i + c_i^{n_i}, & \underline{p}_i^{n_i} \le p_i \le \overline{p}_i^{n_i} \end{cases} \tag{3.56}$$

The integer variables indicate which piece of the cost function p_i occupies. For each piece $k = 1, \ldots, n_i$ of generator i's cost curve, define the new variables ξ_i^k and $\hat{p}_i^k$, and constrain them as

$$\sum_{k=1}^{n_i} \xi_i^k = 1 \tag{3.57}$$

$$\xi_i^k \underline{p}_i^k \le \hat{p}_i^k \le \xi_i^k \overline{p}_i^k, \quad k = 1, \ldots, n_i \tag{3.58}$$

$$\xi_i^k \in \{0, 1\}, \quad k = 1, \ldots, n_i. \tag{3.59}$$

These constraints force the generator to use exactly one fuel source.[7] The objective may then be equivalently written

$$f_i(\hat{p}_i) = \sum_{k=1}^{n_i} a_i^k \left(\hat{p}_i^k \right)^2 + b_i^k \hat{p}_i^k + c_i^k, \tag{3.60}$$

subject to constraints (3.58)-(3.59). The real power output at generator i is then simply

$$p_i = \sum_{k=1}^{n_i} \hat{p}_i^k.$$

---

[7] It is actually only necessary to declare the first $k = n_i - 1$ of ξ_i^k to be binary, because (3.57) implicitly enforces integrality on $\xi_i^{n_i}$. This reduces the amount of binary variables by the number of generators, potentially saving computation time.

Prohibited operating zones, i.e., restricted power output levels, another common non-convexity, can be encoded in this formulation by allowing $\bar{p}_i^j \neq \underline{p}_i^{j+1}$.

The cost function (3.60) and constraints (3.57)-(3.59) can be incorporated into any optimal power flow routine. As discussed in Section 2.2.2, this may only be computationally practical with linear and SOC models, because MISDP is currently undeveloped.

3.5.4   Polyhedral relaxation of the second-order cone

Integer constraints like those above in Section 3.5.3 can dramatically decrease a model's tractability. While MISOCP is gaining ground, it is still far behind MILP, and it remains to be seen if it will ever achieve the same level of maturity. As such, it may sometimes be advantageous to further relax MISOCPs to MILPs even if the underlying continuous problems are of similar difficulty due to the resulting MILP's additional variables and constraints. The methodology of endnote [76] enables us to do exactly this, and to arbitrary accuracy. We refer to it as a *polyhedral relaxation* to distinguish from linear lift-and-project relaxations.

Given the SOC constraint

$$x_1^2 + x_2^2 \leq x_3^2, \tag{3.61}$$

the level n polyhedral relaxation is given by Feasible Set 3.18.

FEASIBLE SET 3.18 (Polyhedral relaxation of an SOC constraint)

$$y_0 \geq |x_1|$$

$$z_0 \geq |x_2|$$

$$y_i = \cos\left(\frac{\pi}{2^{i+1}}\right) y_{i-1} + \sin\left(\frac{\pi}{2^{i+1}}\right) z_{i-1}, \quad i = 1, \ldots, n$$

$$z_i \geq \left|\cos\left(\frac{\pi}{2^{i+1}}\right) z_{i-1} - \sin\left(\frac{\pi}{2^{i+1}}\right) y_{i-1}\right|, \quad i = 1, \ldots, n$$

$$y_n \leq x_3$$

$$z_n \leq \tan\left(\frac{\pi}{2^{n+1}}\right) y_n$$

As $n \to \infty$ and sometimes at finite values of n, the relaxation becomes exact. Any SOC constraint may be decomposed into a collection of constraints in the form of (3.61). For example, by introducing the auxiliary variables Y_1, Y_2, and Y_3, we may write

$$W_{ij}^2 + W_{i+n,j}^2 \leq W_{ii} W_{jj},$$

the SOC constraint from the voltage-polar relaxation of Section 3.3.2, as

$$W_{ij}^2 + W_{i+n,j}^2 \leq Y_1^2$$
$$Y_1^2 + Y_2^2 \leq Y_3^2$$

$$W_{ii} - W_{jj} = 2Y_2$$
$$W_{ii} + W_{jj} = 2Y_3.$$

The relaxation may then be applied to the first two constraints.

3.6   Summary

We have now introduced optimal power flow and a small horde of its convex approximations, assembled roughly in order of decreasing accuracy in Table 3.1. For many years, the linear power flow approximations were the only convex options. The more recent SD and SOC relaxations retain both convexity and many physical aspects of steady-state power flow, unlocking a number of improved power flow applications that we explore in subsequent chapters. By virtue of sitting in the convex optimization classes discussed in Section 2.2, all of the approximations in this chapter can be solved in polynomial-time.

It is important to remember that all of the convex models in this section are approximations; except when the conditions of Theorem 3.1 hold, neither the relaxations nor approximations can be expected to produce feasible, let alone physical, solutions. This is not to say that these approaches are not useful. To wit, the nonconvex, single-phase steady-state power flow derived in Section 2.5 is an approximation to unbalanced three-phase, time-varying power flow, which itself is an approximation to the nonsinusoidal true voltage and current waveforms, and so on down to the most basic physical models. As usual, the central trade-off and key design choice in formulating power system optimizations therefore lie in balancing accuracy and tractability. In the remaining chapters, the models in Table 3.1 will serve as a basis for applying this perspective to a wide variety of power system optimization problems.

Table 3.1 Power flow relaxations (top) and approximations (below), roughly ordered by decreasing accuracy.

Feasible Set	Relaxation	Class	Accuracy (decreasing)
3.1 (Exact AC)	Unrelaxed	NLP	Exact
3.6, 3.9	SD relaxation	SDP	Exact in radial networks
3.8, 3.10, 3.13	SD, SOC relaxations	SOCP	Exact in radial networks
3.18	SD, SOC, polyhedral relaxations	LP	Size dependent
3.14		CQCP	No voltage
3.4 (Decoupled)		LP	Linearized, no losses
3.3 (Linearized)		LP	Linearized, no losses, voltage, or reactive power
3.5 (Network flow)		LP	No electrical physics, same as linearized in radial networks

Let us now discuss avenues for improvement. Table 3.1 is incomplete in two respects.

- While powerful, Theorem 3.1 and its generalizations only characterize the performance of the SD and SOC relaxations in radial networks. On the other hand, examples have been found for which the relaxations produce infeasible (but good approximate) solutions [47]. A number of unanswered theoretical questions about them remain, like the improvement of the SD over the SOC relaxation in meshed networks, and the role of planarity, which affects the complexity of many graph algorithms and is a common trait among power systems.
- By construction, higher order, more accurate linear, SOC, and SD power flow relaxations exist; for example, higher order moment-based relaxations of optimal power flow are investigated in endnote [77]. Do such stronger relaxations offer enough accuracy improvement to justify their increased number of constraints and variables?

From a more applied perspective, much remains to be said about the relaxations' compatibility with further modeling details. For instance, in endnote [78], a semidefinite relaxation analogous to that in Section 3.3 is constructed for unbalanced three-phase power systems, i.e., without the assumptions of Section 2.5.2. To what extent can the relaxations meld with similar increases in resolution?

There are even more fundamental aspects of power flow that we don't fully understand. In particular, we still lack theoretical characterizations of the existence and uniqueness of load flow solutions. For all we currently know, there could exist far more powerful, as yet undiscovered power flow formulations of an entirely different character than those we have discussed.

Problems

3.1   Show that linearized power flow and network flow (Feasible Sets 3.3 and 3.5) are equivalent in radial networks.

3.2   Show that the feasible set of two-commodity network flow is a relaxation of power flow.

3.3   Write the SD relaxation in the standard form given in Section 2.2.2.

3.4   Give a testable condition under which Theorem 3.1 holds when $\underline{p}_i$ and $\underline{q}_i$ are finite.

3.5   Show that Theorem 3.1 holds when $\bar{s}_{ij}$ is finite.

3.6   Construct a three-node network for which the SD relaxation is inexact.

3.7   Construct a three-node network for which the SD relaxation is exact but the SOC relaxation is not.

3.8   Construct a three-node network for which the SOC relaxation is exact.

3.9   Write the SOC constraint (3.22) explicitly in voltage-rectangular coordinates, i.e., when

$$Q = \begin{bmatrix} I & I \\ -I & I \end{bmatrix}.$$

3.10   Show that the exact branch flow model is equivalent to Feasible Set 3.1.

3.11   Show that the SOC branch flow relaxation, Feasible Set 3.13, is equivalent to the other SOC power flow relaxations, e.g., Feasible Set 3.8.

References

[1] J. Jonnes, *Empires of Light: Edison, Tesla, Westinghouse, and the Race to Electrify the World*. Random House, 2004.

[2] R. Munson, *From Edison to Enron: the Business of Power and What It Means for the Future of Electricity*. Praeger Publishers, 2005.

[3] J. Carpentier, "Contribution to the economic dispatch problem," *Bulletin de la Societe Francoise des Electriciens*, vol. 3, no. 8, pp. 431–447, 1962.

[4] W. Tinney and C. Hart, "Power flow solution by Newton's method," *IEEE Transactions on Power Apparatus and Systems*, vol. 86, no. 11, pp. 1449–1460, Nov. 1967.

[5] H. Dommel and W. Tinney, "Optimal power flow solutions," *IEEE Transactions on Power Apparatus and Systems*, vol. 87, no. 10, pp. 1866–1876, Oct. 1968.

[6] B. Stott and O. Alsac, "Fast decoupled load flow," *IEEE Transactions on Power Apparatus and Systems*, vol. 93, no. 3, pp. 859–869, May 1974.

[7] D. Sun, B. Ashley, B. Brewer, A. Hughes, and W. Tinney, "Optimal power flow by Newton approach," *IEEE Transactions on Power Apparatus and Systems*, vol. 103, no. 10, pp. 2864–2880, Oct. 1984.

[8] O. Alsac, J. Bright, M. Prais, and B. Stott, "Further developments in LP-based optimal power flow," *IEEE Transactions on Power Systems*, vol. 5, no. 3, pp. 697–711, Aug. 1990.

[9] S. Granville, "Optimal reactive dispatch through interior point methods," *IEEE Transactions on Power Systems*, vol. 9, no. 1, pp. 136–146, Feb. 1994.

[10] Y.-C. Wu, A. Debs, and R. Marsten, "A direct nonlinear predictor-corrector primal-dual interior point algorithm for optimal power flows," *IEEE Transactions on Power Systems*, vol. 9, no. 2, pp. 876–883, May 1994.

[11] G. Torres and V. Quintana, "An interior-point method for nonlinear optimal power flow using voltage rectangular coordinates," *IEEE Transactions on Power Systems*, vol. 13, no. 4, pp. 1211–1218, Nov. 1998.

[12] R. Jabr, A. H. Coonick, and B. J. Cory, "A primal-dual interior point method for optimal power flow dispatching," *IEEE Power Engineering Review*, vol. 22, no. 7, p. 55, July 2002.

[13] M. Huneault and F. Galiana, "A survey of the optimal power flow literature," *IEEE Transactions on Power Systems*, vol. 6, no. 2, pp. 762–770, 1991.

[14] J. Momoh, R. Adapa, and M. El-Hawary, "A review of selected optimal power flow literature to 1993. I. Nonlinear and quadratic programming approaches," *IEEE Transactions on Power Systems*, vol. 14, no. 1, pp. 96–104, 1999.

[15] J. Momoh, M. El-Hawary, and R. Adapa, "A review of selected optimal power flow literature to 1993. II. Newton, linear programming and interior point methods," *IEEE Transactions on Power Systems*, vol. 14, no. 1, pp. 105–111, Feb. 1999.

[16] L. Krumm, "Summary of the gradient method for optimizing the power flow in an interconnected power system (in Russian)," *Isv. Acad. Nauk USSR*, vol. 3, pp. 3–16, 1965.

[17] R. Jabr, "Radial distribution load flow using conic programming," *IEEE Transactions on Power Systems*, vol. 21, no. 3, pp. 1458–1459, Aug. 2006.

[18] X. Bai, H. Wei, K. Fujisawa, and Y. Wang, "Semidefinite programming for optimal power flow problems," *International Journal of Electrical Power and Energy Systems*, vol. 30, no. 6-7, pp. 383–392, 2008.

[19] J. Lavaei and S. Low, "Zero duality gap in optimal power flow problem," *IEEE Transactions on Power Systems*, vol. 27, no. 1, pp. 92–107, Feb. 2012.

[20] B. Zhang and D. Tse, "Geometry of injection regions of power networks," *IEEE Transactions on Power Systems*, vol. 28, no. 2, pp. 788–797, 2013.

[21] D. Molzahn, J. Holzer, B. Lesieutre, and C. DeMarco, "Implementation of a large-scale optimal power flow solver based on semidefinite programming," *IEEE Transactions on Power Systems*, vol. 28, no. 4, pp. 3987–3998, 2013.

[22] S. Low, "Convex relaxation of optimal power flow – Part I: Formulations and equivalence," *IEEE Transactions on Control of Network Systems*, vol. 1, no. 1, pp. 15–27, March 2014.

[23] ——, "Convex relaxation of optimal power flow – Part II: Exactness," *IEEE Transactions on Control of Network Systems*, pp. 15–27, June 2014.

[24] S. Borenstein, M. Jaske, and A. Rosenfeld, "Dynamic pricing, advanced metering, and demand response in electricity markets," UC Berkeley: Center for the Study of Energy Markets, Tech. Rep., 2002.

[25] D. Kirschen, "Demand-side view of electricity markets," *IEEE Transactions on Power Systems*, vol. 18, no. 2, pp. 520–527, 2003.

[26] T. Gedra and P. Varaiya, "Markets and pricing for interruptible electric power," *IEEE Transactions on Power Systems*, vol. 8, no. 1, pp. 122–128, 1993.

[27] A. J. Wood and B. F. Wollenberg, *Power Generation, Operation, and Control*, 3rd ed. Wiley, 2013.

[28] M. Shahidehpour, H. Yamin, and Z. Li, *Market Operations in Electric Power Systems (Forecasting, Scheduling, and Risk Management)*. Wiley-IEEE Press, 2002.

[29] S. Stoft, *Power System Economics: Designing Markets for Electricity*. Wiley-IEEE Press, 2002.

[30] H. Wei, H. Sasaki, J. Kubokawa, and R. Yokoyama, "An interior point nonlinear programming for optimal power flow problems with a novel data structure," *IEEE Transactions on Power Systems*, vol. 13, no. 3, pp. 870–877, 1998.

[31] M. Bazaraa, H. Sherali, and C. Shetty, *Nonlinear Programming Theory and Algorithms*. New York: John Wiley, 1993.

[32] S. Wright, *Primal-Dual Interior-Point Methods*. Society for Industrial and Applied Mathematics, 1997.

[33] J. Nocedal and S. J. Wright, *Numerical Optimization*. Springer, 2000.

[34] D. P. Bertsekas, *Nonlinear Programming*. Athena Scientific, 1999.

[35] S. Boyd and L. Vandenberghe, *Convex Optimization*. New York: Cambridge University Press, 2004.

[36] Y. Nesterov and A. Nemirovski, *Interior Point Polynomial Methods in Convex Programming*. SIAM Studies in Applied Mathematics, 1994, vol. 13.

[37] M. Bollen, *Understanding Power Quality Problems: Voltage Sags and Interruptions*. IEEE Press, 2000.

[38] R. Dugan, S. Santoso, M. McGranaghan, and H. Beaty, *Electrical Power Systems Quality*, ser. McGraw-Hill Professional Engineering. McGraw-Hill, 2003.

[39] L. R. Ford Jr. and D. R. Fulkerson, *Flows in Networks*. Princeton University Press, 1962.

[40] R. K. Ahuja, T. L. Magnanti, and J. B. Orlin, *Network Flows: Theory, Algorithms, and Applications*. Upper Saddle River, NJ: Prentice-Hall, 1993.

[41] M. S. Bazaraa, J. J. Jarvis, and H. D. Sherali, *Linear Programming and Network Flows*. Wiley-Interscience, 2004.

[42] R. A. Horn and C. R. Johnson, *Matrix Analysis*. Cambridge University Press, 1990.

[43] R. Louca, P. Seiler, and E. Bitar, "A rank minimization algorithm to enhance semidefinite relaxations of optimal power flow," in *2012 51st Annual Allerton Conference on Communication, Control, and Computing (Allerton)*, Sept. 2013.

[44] R. Jabr, "Exploiting sparsity in SDP relaxations of the OPF problem," *IEEE Transactions on Power Systems*, vol. 27, no. 2, pp. 1138–1139, 2012.

[45] A. Lam, B. Zhang, and D. Tse, "Distributed algorithms for optimal power flow problem," in *2012 IEEE 51st Annual Conference on Decision and Control (CDC)*, 2012, pp. 430–437.

[46] M. S. Andersen, A. Hansson, and L. Vandenberghe, "Reduced-complexity semidefinite relaxations of optimal power flow problems," *IEEE Transactions on Power Systems*, 2014.

[47] B. Lesieutre, D. Molzahn, A. Borden, and C. DeMarco, "Examining the limits of the application of semidefinite programming to power flow problems," in *2011 49th Annual Allerton Conference on Communication, Control, and Computing (Allerton)*, Sept. 2011, pp. 1492–1499.

[48] S. Kim and M. Kojima, "Exact solutions of some nonconvex quadratic optimization problems via SDP and SOCP relaxations," *Computational Optimization and Applications*, vol. 26, pp. 143–154, 2003.

[49] M. X. Goemans and D. P. Williamson, "Approximation algorithms for max-3-cut and other problems via complex semidefinite programming," *Journal of Computer and System Sciences*, vol. 68, no. 2, pp. 442–470, 2004.

[50] M. Baran and F. Wu, "Network reconfiguration in distribution systems for loss reduction and load balancing," *IEEE Transactions on Power Delivery*, vol. 4, no. 2, pp. 1401–1407, Apr. 1989.

[51] ——, "Optimal capacitor placement on radial distribution systems," *IEEE Transactions on Power Delivery*, vol. 4, no. 1, pp. 725–734, Jan. 1989.

[52] J. A. Taylor and F. S. Hover, "Convex models of distribution system reconfiguration," *IEEE Transactions on Power Systems*, vol. 27, no. 3, pp. 1407–1413, Aug. 2012.

[53] M. Farivar and S. Low, "Branch flow model: Relaxations and convexification – Part I," *IEEE Transactions on Power Systems*, vol. 28, no. 3, pp. 2554–2564, 2013.

[54] ——, "Branch flow model: Relaxations and convexification – Part II," *IEEE Transactions on Power Systems*, vol. 28, no. 3, pp. 2565–2572, 2013.

[55] J. A. Taylor and F. S. Hover, "Conic AC transmission system planning," *IEEE Transactions on Power Systems*, vol. 28, no. 2, pp. 952–959, 2013.

[56] E. J. Candès, T. Strohmer, and V. Voroninski, "PhaseLift: Exact and stable signal recovery from magnitude measurements via convex programming," *Communications on Pure and Applied Mathematics*, vol. 66, no. 8, pp. 1241–1274, 2013.

[57] Y. Tamura, H. Mori, and S. Iwamoto, "Relationship between voltage instability and multiple load flow solutions in electric power systems," *IEEE Transactions on Power Apparatus and Systems*, vol. 102, no. 5, pp. 1115–1125, 1983.

[58] S. Granville, J. Mello, and A. Melo, "Application of interior point methods to power flow unsolvability," *IEEE Transactions on Power Systems*, vol. 11, no. 2, pp. 1096–1103, 1996.

[59] H.-D. Chiang and M. Baran, "On the existence and uniqueness of load flow solution for radial distribution power networks," *IEEE Transactions on Circuits and Systems*, vol. 37, no. 3, pp. 410–416, 1990.

[60] F. R. K. Chung, *Spectral Graph Theory (CBMS Regional Conference Series in Mathematics, No. 92)*. American Mathematical Society, February 1997.

[61] A. Ben-Israel and T. Greville, *Generalized Inverses: Theory and Applications*, ser. CMS Books in Mathematics. Springer, 2003.

[62] M. Baran and N. Mahajan, "DC distribution for industrial systems: Opportunities and challenges," *IEEE Transactions on Industry Applications*, vol. 39, no. 6, pp. 1596–1601, Dec. 2003.

[63] D. Salomonsson and A. Sannino, "Low-voltage DC distribution system for commercial power systems with sensitive electronic loads," *IEEE Transactions on Power Delivery*, vol. 22, no. 3, pp. 1620–1627, July 2007.

[64] R. Lasseter, "Microgrids," in *Power Engineering Society Winter Meeting, 2002*, vol. 1, 2002, pp. 305–308.

[65] J. Guerrero, J. Vasquez, J. Matas, L. de Vicuña, and M. Castilla, "Hierarchical control of droop-controlled AC and DC microgrids – A general approach toward standardization," *IEEE Transactions on Industrial Electronics*, vol. 58, no. 1, pp. 158–172, 2011.

[66] N. Flourentzou, V. Agelidis, and G. Demetriades, "VSC-based HVDC power transmission systems: An overview," *IEEE Transactions on Power Electronics*, vol. 24, no. 3, pp. 592–602, March 2009.

[67] R. Erickson and D. Maksimovic, *Fundamentals of Power Electronics*. Springer, 2001.

[68] D. Hammerstrom, "AC versus DC distribution systems: Did we get it right?" in *Power Engineering Society General Meeting*, June 2007, pp. 1–5.

[69] L. Gan and S. Low, "Optimal power flow in DC networks," in *2013 IEEE 52nd Annual Conference on Decision and Control (CDC)*, 2013.

[70] J. Lamont and J. Fu, "Cost analysis of reactive power support," *IEEE Transactions on Power Systems*, vol. 14, no. 3, pp. 890–898, Aug. 1999.

[71] K. Bhattacharya and J. Zhong, "Reactive power as an ancillary service," *IEEE Transactions on Power Systems*, vol. 16, no. 2, pp. 294–300, May 2001.

[72] P. Kundur, *Power System Stability and Control*. McGraw-Hill Professional, 1994.

[73] C. Lin and G. Viviani, "Hierarchical economic dispatch for piecewise quadratic cost functions," *IEEE Transactions on Power Apparatus and Systems*, vol. 103, no. 6, pp. 1170–1175, June 1984.

[74] F. Lee and A. Breipohl, "Reserve constrained economic dispatch with prohibited operating zones," *IEEE Transactions on Power Systems*, vol. 8, no. 1, pp. 246–254, Feb. 1993.

[75] D. Walters and G. Sheble, "Genetic algorithm solution of economic dispatch with valve point loading," *IEEE Transactions on Power Systems*, vol. 8, no. 3, pp. 1325–1332, Aug. 1993.

[76] A. Ben-Tal and A. Nemirovski, "On polyhedral approximations of the second-order cone," *Mathematics of Operations Research*, vol. 26, no. 2, pp. 193–205, 2001.

[77] D. K. Molzahn and I. A. Hiskens, "Moment-based relaxation of the optimal power flow problem," in *Proc. 18th Power Systems Computation Conference (PSCC)*, Aug. 2014.

[78] E. Dall'Anese, G. Giannakis, and B. Wollenberg, "Optimization of unbalanced power distribution networks via semidefinite relaxation," in *North American Power Symposium*, Sept. 2012, pp. 1–6.

4 System operation

The day-to-day and long-term operation of power systems is made up of a sprawling collection of engineering and economic tasks that subsumes most of the contents of this book. This chapter does not address the full scope of power system operation but rather a handful of relevant problems that are amenable to convex optimization. As will be seen, many of these problems consist of details layered on top of the optimal power flow formulations of Chapter 3. The interested reader is referred to Wood and Wollenberg [1] for broad coverage of power system operation.

Of particular relevance to this chapter is Theorem 3.1 of Section 3.3.1, which guarantees that under certain conditions, the SOC optimal power flow relaxation is exact in radial networks. It is therefore a great boon that distribution systems, the portions of power systems that convey low-voltage electricity from substations to end users, are almost always operated radially. Until recently, distribution systems were essentially passive, predictable energy consumers. The shift from fuel-based centralized power plants to distributed and renewable generation and the new, active role of loads like electric vehicles and smart buildings are transforming distribution systems into highly actuated, potentially volatile consumers *and* producers of electric power. Consequently, many of the formulations in this chapter are both highly tractable and relevant when specialized to radial networks. This is especially true in Section 4.4, which deals exclusively with power flow in radial networks.

4.1 Multi-period optimal power flow

Optimal power flow routines are run every few minutes to update device and resource settings in response to the constantly changing conditions of power systems. The dynamic couplings present over these time scales justify linking these routines over successive time periods. This section formulates the *multi-period* optimal power flow and elucidates its application through generator ramp constraints and energy storage.

Essentially, multi-period optimal power flow is just a sequence of ordinary optimal power flow routines strung together by dynamic costs and constraints. This is accomplished by creating a set of power flow variables for each time period, indexed $t \in \{1, \ldots, T\}$ and evenly spaced by intervals of duration Δ. These variables are coupled through a cost function,

$$f\left(v^1, \ldots, v^T, p^1, \ldots, p^T, q^1, \ldots, q^T, \Delta\right), \tag{4.1}$$

and dynamic constraints,

$$g_i^t \left(p^t, p^{t-1}, q^t, q^{t-1}, z^t, z^{t-1}, \Delta \right) \leq 0, \tag{4.2}$$

where z^t is a vector of auxiliary variables.[1] As long as f and each g_i^t are convex, the corresponding multi-period optimal power flow is also convex. Note that if (4.2) implies an equality constraint, it must be affine to be convex, i.e., of the form

$$\begin{bmatrix} p^t \\ q^t \\ z^t \end{bmatrix} = A_i^t \begin{bmatrix} p^{t-1} \\ q^{t-1} \\ z^{t-1} \end{bmatrix} + b_i^t. \tag{4.3}$$

The multi-period optimal power flow is composed of (4.1), (4.2), and T copies of a feasible set of power flows from Chapter 3. For example, if f and each g_i^t were linear, the linearized power flow (Feasible Set 3.3) would result in an LP. Similarly, using the SD relaxation (e.g., Feasible Set 3.9) at each time period would accommodate more general forms of f and g_i^t expressible via SDP. Clearly, computational scalability is a concern because the number of variables and constraints in standard optimal power flow have increased by a factor of T.

Observe that despite being dynamically coupled, multi-period optimal power flow is steady-state like ordinary optimal power flow, as discussed in Section 3.1. Faster, often nonlinear dynamics such as the swing equations (Section 4.2.1) are typically orders of magnitude faster than optimal power flow repetitions. As indicated by the two applications in this section, real power is usually the only quantity dynamically coupled over steady-state time scales.

We distinguish between *policy* and *trajectory* solutions. Informally, call x the *state* and u the *control* variables; u is chosen to affect x, but x might also be subject to random disturbances. For instance, x could be power imbalances or voltage angles across the system, and u nodal power injections. Control policies choose u as a function of x, i.e.,

$$u^t = \mu^t \left(x^t \right),$$

where μ^t is the control policy at time t. Here, little is assumed about *actual* future values of x because real observations are used online.

Trajectory-based approaches do not distinguish between state and control variables but rather treat both as optimization variables within one large problem. A trajectory is composed of sequences $x^1, \ldots, x^T$ and $u^1, \ldots, u^T$. Consequently, an optimal control trajectory loses validity if the system deviates from the associated optimal state trajectory. Policies are more useful than trajectories because they apply to all states rather than those on a single path, and they can incorporate new information as it becomes available. Optimal policies are also usually far more difficult to obtain analytically and computationally.

---

[1] The dynamic constraints only couple pairs of consecutive periods for concision and because the two examples in this section, ramp constraints and storage, are of this form. However, the formulation straightforwardly handles constraints over any subset of periods, as appear in unit commitment in Section 4.3.

As written above, multi-period optimal power flow produces trajectories and not policies, but it is able to accommodate essentially all ordinary optimal power flow modeling. On the other hand, current techniques for obtaining policies either assume considerably simplified physics such as in linear quadratic regulation (Section 4.2.2) and *risk-limiting dispatch* [2, 3], or they apply only to special cases like a single energy storage (Section 4.1.2). However, in scenarios where uncertainty is substantial, for example hour-ahead resource scheduling with intermittent renewable generation, the flexibility of a policy can outweigh the benefits of detailed modeling. Approximate approaches for obtaining suboptimal policies for detailed models are promising in such scenarios, cf. [4]. Section 4.1.3 briefly discusses how to use routines for generating trajectories as policies via model predictive control.

4.1.1 Ramp constraints

A ramp is a change in a resource's real power input or output, or, mathematically, its derivative. *Uncontrolled ramping* such as a sudden doubling in a wind farm's output is considered a disturbance [5], while the capability of resources like combined cycle gas turbines and energy storage to undergo *fast ramping* is a highly valued power system service [6]. A ramp constraint on a single resource's real power output is shown in Figure 4.1.

Suppose $\underline{r}_i^t$ and $\bar{r}_i^t$ are respectively the minimum and maximum allowable ramp rates between periods $t - 1$ and t. In this case, (4.2) takes the form

$$\underline{r}_i^t \le p_i^t - p_i^{t-1} \le \bar{r}_i^t. \tag{4.4}$$

To penalize ramping in the objective, (4.1) would contain terms proportional to $\left(p_i^t - p_i^{t-1}\right)^2$ for each resource for which excessive ramping is undesirable. The above ramp constraint is incorporated into a multi-period optimal power flow in the following example.

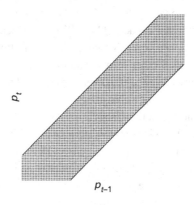

p_t

p_{t-1}

Fig. 4.1   The feasible region described by a ramp constraint.

Example 4.1 *Multi-period linearized power flow with ramp constraints.* Suppose we would like to minimize generator fuel and ramping costs, $f_i^t\left(p_i^t\right)$ and $\left(p_i^t - p_i^{t-1}\right)^2$, respectively, over the time periods $t \in \{1, \dots, T\}$. The linearized power flow version of this problem is given by:

$$\underset{p,\theta}{\text{minimize}} \ \sum_t \sum_i f_i^t\left(p_i^t\right) + \left(p_i^t - p_i^{t-1}\right)^2$$

$$\text{subject to} \ \ p_{ij}^t = b_{ij}\left(\theta_i^t - \theta_j^t\right)$$

$$\sum_j p_{ij}^t = p_i^t$$

$$\underline{p}_i^t \leq p_i^t \leq \overline{p}_i^t$$

$$\left|p_{ij}^t\right| \leq \overline{s}_{ij}$$

$$\underline{r}_i^t \leq p_i^t - p_i^{t-1} \leq \overline{r}_i^t$$

A solution to this problem is composed of sequences of nodal exchanges of real power, p_i^t, and voltage angles, θ_i^t. Because it is a trajectory and not a policy, such a solution does not anticipate the future availability of new information.

4.1.2   Energy storage and inventory control

Renewable energy sources displace conventional, polluting power sources, but they increase the need for reserves and regulation traditionally provided by those same resources, potentially negating the environmental and economic benefits of renewables. This has created a niche for pure sources of regulation and reserves, namely energy storage. Unconstrained by the need to generate power, technologies such as batteries, flywheels, and supercapacitors are capable of providing more precise regulation and reserves with minuscule lead time. Of course, pumped hydro energy storage has been a mainstay of power systems since the early 1900s and makes up the overwhelming majority of installed capacity [1, 7]. While conceptually similar, pumped hydro energy storage is subject to severe geographical restrictions and costs that make new projects less pragmatic than other forms. The interested reader is referred to endnotes [8, 9] for surveys of energy storage in power systems and to endnote [10] for broader coverage of energy storage and renewable energy.

A second non-generator source of regulation and reserves is demand response, in which electrical loads allow system operators to modulate their consumption for a financial reward [11, 12]. One reason demand response programs are attractive to the power industry is that, unlike energy storage, they require minimal additional infrastructural investment. Although a demand response program may have hundreds of thousands of participants, the aggregation can be approximated as energy storage with time-varying parameters [13, 14], which is commensurate with the resolution of optimal power flow. As such, both resources can be described by the following storage model.

Denote a storage's stored energy or *state of charge* at period t by s_i^t. s_i^t must stay between zero and the storage's maximum energy capacity, $\overline{c}_i^t$. The grid-side power injection and extractions are respectively denoted u_i^{it} and u_i^{et}, the difference of which is the net power transacted, p_i^t. p_i^t obeys a real power constraint like (3.6) but which is more conveniently enforced on u_i^{it} and u_i^{et}. Energy is dissipated while it sits inside ("leakage") or when it is moved into or out of storage, which we respectively model with the scaling factors α_i^t, η_i^{et}, and η_i^{it}, each of which lies in the interval $(0, 1]$. The feasible set for a single energy storage is given below.

FEASIBLE SET 4.1 (Linear energy storage)

$$s_i^t = \alpha_i^t s_i^{t-1} + \Delta \left(\eta_i^{et} u_i^{et-1} - u_i^{it-1}/\eta_{it}^i \right) \tag{4.5}$$

$$0 \le s_i^t \le \overline{c}_i^t \tag{4.6}$$

$$p_i^t = u_i^{it} - u_i^{et} \tag{4.7}$$

$$0 \le u_i^{et} \le -\underline{p}_i^t \tag{4.8}$$

$$0 \le u_i^{it} \le \overline{p}_i^t \tag{4.9}$$

Constraint (4.5) is the storage-specific form of (4.3). Other non-dynamic constraints can be straightforwardly added to each individual period's constraints in the multi-period feasible set. Note that the above feasible set does not prohibit storage from simultaneously injecting and extracting power. In fact, it is rarely necessary to impose such a constraint because it is rarely optimal to dissipate energy through conversion losses. If it is, such an action may be physically meaningful, for example, in the case of negative power prices that incentivize energy disposal. Nevertheless, a *complementarity* constraint may be enforced on u_i^{it} and u_i^{et} using binary variables by replacing the last two constraints in Feasible Set 4.1 with the following.

$$0 \le u_i^{et} \le -\underline{p}_i^t \xi_i^t$$
$$0 \le u_i^{it} \le \overline{p}_i^t \left(1 - \xi_i^t \right)$$
$$\xi_i^t \in \{0, 1\}.$$

These binary variables contribute substantial computational cost, potentially to the extent that we can only attain suboptimal solutions that are inferior to optimal solutions without the above constraints.

Unlike generators, which are often directly connected to AC power systems, energy storage is usually interfaced through power electronic converters [15, 16]. These converters afford a great deal of flexibility to power system operations (albeit, in exchange for some undesirable harmonics) because they can inject and extract reactive power, potentially subsuming the functionality of purely reactive power sources like capacitor banks and static VAR and synchronous compensators. Detailed modeling of power electronic devices is well beyond the scope of this book. The following apparent power constraint is sufficient for the purposes of multi-period optimal power flow:

$$\left(p_i^t\right)^2 + \left(q_i^t\right)^2 \le \left(\bar{s}_i^t\right)^2.$$

Like the constraint transmission capacity imposes on apparent power flows, (3.8), this is a convex quadratic constraint, which can be incorporated without modification into SOC or SD formulations. A linear approximation such as the following outer approximation is necessary for use in purely linear models:

$$\left|p_i^t\right| + \left|q_i^t\right| \le \sqrt{2}\bar{s}_i^t$$
$$\left|p_i^t\right| \le \bar{s}_i^t$$
$$\left|q_i^t\right| \le \bar{s}_i^t$$

Example 4.2   *Fast ramping energy storage in a distribution system.* As discussed in the beginning of this chapter, distribution systems are home to many new renewable power sources, energy storages, and sensors. The resulting enhanced controllability justifies multi-period optimal power flow-based operation of distribution systems. Because they are usually radial, we can expect the SOC relaxation of Section 3.3 to perform well and to be exact if the conditions of Theorem 3.1 are met. This example combines various elements of this section with the SOC branch flow relaxation of Section 3.3.3 in a radial distribution system.

Suppose that power can be produced at a subset of nodes, $\mathcal{G}$, and that in addition to generation costs, there is a convex ramping cost $f_i\left(p_i^t, p_i^{t-1}\right)$. A susbset of nodes, $\mathcal{S}$, has energy storage, each of which is constrained by Feasible Set 4.1. Both generators and storage are constrained by apparent power and ramp limits; note that here apparent power limits are used to approximate the reactive power capability curve of Section 3.5.2. We compose these features in the following multi-period optimal power flow routine:

$$\underset{p,q,V}{\text{minimize}} \quad \sum_t \sum_{i \in \mathcal{G}} f_i\left(p_i^t\right) + f_i\left(p_i^t, p_i^{t-1}\right)$$

subject to   A copy of Feasible Set 3.13 for each time period

[SOC power flow relaxation]

A copy of Feasible Set 4.1 for each time period and each $i \in \mathcal{S}$

[Energy storage constraints]

$$\left(p_i^t\right)^2 + \left(q_i^t\right)^2 \le \left(\bar{s}_i^t\right)^2 \quad \text{for each } i \in \mathcal{S} \cup \mathcal{G}$$
$$\underline{r}_i^t \le p_i^t - p_i^{t-1} \le \bar{r}_i^t \quad \text{for each } i \in \mathcal{S} \cup \mathcal{G}$$

In this feasible set, the branch flow relaxation has been replicated in each time period and coupled sequentially by the storage and ramp constraints. The resulting optimization problem is an SOC. Indeed, every constraint is linear with the exception of the quadratically constrained transmission capacity and apparent power limits and the SOC constraint in the branch flow model.

Theorem 3.1 tells us that nothing can be gained from using an SD instead of an SOC relaxation because the network is radial. While the theorem does not perfectly apply because of the apparent power flow limits and lower nodal power limits, we take it as evidence that this model contains a very good approximate if not exact power flow representation.

Inventory control policies

As discussed at the start of this section, the above formulations are only useful if the system stays on or near the optimal trajectory. This is a reasonable assumption if the system is accurately modeled and has small random disturbances relative to the time horizon. Absent these conditions, control policies are far superior because they return a decision for any system state regardless of the particular trajectory. Because absorbing random disturbances from renewable sources is a primary function of energy storage, it is highly appropriate to seek policies for energy storage. Unfortunately, control policies are usually more costly then trajectories and are often only obtainable through computationally intensive dynamic programming approaches. The most notable exception is linear systems, for which the optimal *linear quadratic regulator* is analytically tractable, which we discuss in Section 4.2.2.

Inventory control is a lesser known scenario for which optimal policies are also easy to obtain, which was originally motivated by businesses stocking goods under random demand, cf. [17, 18]. Here, we follow the formulation of [19] to show that optimally scheduling a single energy storage is almost perfectly analogous to inventory control.

Because we are considering a single storage, here we drop the subscript i and assume that each period has unit duration, i.e., $\Delta = 1$. We also now assume that injection and extraction are perfectly efficient, $\eta^{et} = \eta^{it} = 1$, and that there are no power limits. This enables u^{et} and u^{it} to be collapsed into a single quantity, u^t. Suppose δ^t is a random amount of energy seen by the energy storage at time t, for example from load variation or a local renewable source's intermittency. Nothing is assumed about the probability distribution of δ^t.

Our goal is to find a policy, $\mu^t\left(s^t\right)$, which outputs the optimal power exchange, u^t, given the state of charge s^t. Define the saturation function

$$\mathrm{sat}^t(x, y) = \max\left\{\min\left\{x, \overline{c}^t - y\right\}, -y\right\}.$$

This truncates the quantity x at upper and lower limits determined by the capacity and the dummy variable y. The state of charge, formerly described by (4.5), now evolves as

$$s^{t+1} = \alpha^t s^t + u^t + \mathrm{sat}^t\left(\delta^t, \alpha^t s^t + u^t\right). \tag{4.10}$$

Suppose that the goal is to solve

$$\underset{u^t}{\mathrm{minimize}}\ \mathbb{E}\left[\sum_t g^t\left(\delta^t, \alpha^t s^t + u^t\right) + \lambda^t u^t\right] \tag{4.11}$$

subject to $0 \leq s^t \leq \bar{c}^t$ and (4.10), where $\mathbb{E}$ denotes the expectation over δ^t and λ^t is the price of power at time t. For example,

$$g^t(\delta^t, \alpha^t s^t + u^t) = \left| \delta^t - \text{sat}^t \left(\delta^t, \alpha^t s^t + u^t \right) \right|$$

represents the portion of the random energy not captured by storage at time t, also known as the *spillover*. The term $\lambda^t u^t$ encodes profits to be made by buying and selling at respectively low and high prices, also known as load shifting and intertemporal arbitrage.

We solve this problem using dynamic programming; the reader is referred to endnote [17] for broad coverage of dynamic programming. Let $J^t \left(s^t \right)$ denote the value function at time t, i.e., the sum of the terms in (4.11) with time index greater than or equal to t given that the current state is s^t. The dynamic programming recursion is given by

$$J^t \left(s^t \right) = \min_{u^t} \; \mathbb{E} \Big[g^t \left(\delta^t, \alpha^t s^t + u^t \right) + \lambda^t u^t +$$
$$J^{t+1} \left(\alpha^t s^t + u^t + \text{sat}^t \left(\delta^t, \alpha^t s^t + u^t \right) \right) \Big]$$
$$\text{subject to} \quad 0 \leq \alpha^t s^t + u^t \leq \bar{c}^t.$$

Often, dynamic programs like this can only be solved by discretizing s^t and u^t and enumerating trajectories. The key step that makes inventory control problems analytically tractable is substituting a new variable y^t for each instance of $\alpha^t s^t + u^t$. y^t represents the state of charge after the control action but immediately prior to the realization of δ^t. We can then rewrite the above dynamic program as

$$J^t(s^t) = -\lambda^t \alpha^t s^t + \min_{y^t} \; \mathbb{E} \Big[g^t \left(\delta^t, y^t \right) + \lambda^t y^t + J^{t+1} \left(y^t + \text{sat}^t \left(\delta^t, y^t \right) \right) \Big]$$
$$\text{subject to} \quad 0 \leq y^t \leq \bar{c}^t.$$

Notice now that the state, s^t, does not appear within the minimization over y^t. We solidify this insight by defining the intermediary function

$$G^t(y^t) = \mathbb{E} \Big[g^t \left(\delta^t, y^t \right) + \lambda^t y^t + J^{t+1} \left(y^t + \text{sat}^t \left(\delta^t, y^t \right) \right) \Big]$$

and denote its minimizer by

$$z^t = \underset{0 \leq y^t \leq \bar{c}^t}{\text{argmin}} \; G^t \left(y^t \right).$$

Reversing the substitution, we can thus write the value function and corresponding optimal control policy

$$J^t(s^t) = G^t \left(z^t \right) - \lambda^t \alpha^t s^t,$$
$$\mu^t(s^t) = z^t - \alpha^t s^t.$$

The value function and optimal control policy are affine functions of s^t, implying that z^t need only be found once for each time t rather than for each possible state s^t. z^t may be solved for by minimizing over $G^t \left(y^t \right)$ at the final period and then iterating backwards in

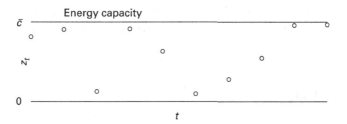

Fig. 4.2   An inventory control policy's optimal set-points, z^t, for storage with constant energy capacity $\bar{c}$.

time. Intuitively, z^t is an optimal sequence of set-points: regardless of the actual past or present state of charge, z^t is the optimal place to be at time t. An example sequence is illustrated in Figure 4.2.

The inventory control formulation rests on multiple idealizations, some of which are severe. Nevertheless, it serves well as a platform for heuristic, suboptimal policies that incorporate more realistic details, cf. [19]. For example, power constraints on u^t can be heuristically incorporated by truncating the idealized policy:

$$\mu_h^t\left(s^t\right) = \begin{cases} \underline{p}_i^t & \mu^t(s^t) < \underline{p}_i^t \\ \overline{p}_i^t & \mu^t(s^t) > \overline{p}_i^t \\ \mu^t\left(s^t\right) & \text{otherwise} \end{cases}.$$

4.1.3   Implementation via model predictive control

Policies such as inventory control are unambiguously implementable because they give instructions for any state. Trajectory solutions, however, usually become less valid with each successive period as the true state diverges from the optimal trajectory due to disturbances and imperfect modeling. *Model predictive control* is a general approach to this issue in which a new optimal trajectory is computed at each time period, but only the current period's control actions are applied. Broad expositions of model predictive control are given by endnotes [20, 21].

A simple model predictive control implementation of multi-period optimal power flow is as follows. Let T_0 be the starting period and T_H be a user-specified time horizon.

1 Solve a multi-period optimal power flow ranging from time $t = T_0, \ldots, T_0 + T_H$.
2 Apply the resulting optimal control variables from period T_0. For example, generator i might be commanded to output $p_i^{T_0}$ and $q_i^{T_0}$ at time T_0.
3 Observe new system information and update the parameters of the multi-period optimal power flow routine. For instance, this may include observing the system state through measurements, running a state estimator, or updating load or renewable forecasts.
4 Increment $T_0 \leftarrow T_0 + 1$, and go to step one.

This type of approach is also commonly referred to as *receding horizon control*. Far more sophisticated model predictive control algorithms than the above exist, including policy and decentralized variants.

4.2   Stability and control

This section departs from steady-state modeling to consider the inertial dynamics of power systems. In regard to the previous section, we are allowing the time between successive periods to approach zero, so that electromechanical physics become significant. This entails using differential rather than difference equations, although this is not an important mathematical distinction. The phenomena we are now interested in occur at relatively fast time scales of under a few seconds and can be critical in determining if a disturbance like a transmission line failure is absorbed or snowballs into a blackout [22, 23]. Such events are often quantified in terms of *frequency instability* and *voltage collapse*, which are defined shortly.

Disturbances are absorbed through *reserves* and *regulation*. Reserves are spare capacity typically provided by energy storage or generators operating on standby; for example, if load suddenly increases or a generator unexpectedly disconnects or *trips*, reserves maintain stability and ensure that the ensuing service disruption is brief or nonexistent. Regulation is continually applied to cancel smaller, faster disturbances such as from faults, load variation, or wind and solar intermittency.

Frequency regulation ensures that the system stays at its nominal voltage frequency, which is 60 Hz in North America and 50 Hz in Europe. *Automatic generation control* provides coarse frequency regulation by updating each generator's real power output every few seconds. Internally, *speed governors* enable most generators to smoothly follow automatic generation signals and damp higher frequency disturbances. Voltage regulation keeps voltage magnitudes within acceptable ranges, which are usually somewhat arbitrarily chosen to be around 0.95 and 1.05 per unit. Voltage regulation is achieved either through excitation in generators or reactive power injections from capacitor banks and static VAR compensators. The approximations of Section 3.2.1 highlight the networked dependence of voltage angle on real power and the strongly local dependence of voltage magnitude on reactive power. Note that this decoupling becomes dubious in distribution systems where resistances account for nontrivial portions of impedances.

The basic model capturing power system dynamics and regulation is known as the *swing equation*. This section presents the basic swing equation and discusses control methods for regulation. Beyond its general importance, an additional motivation for studying the swing equation is its mathematical relationship to the power flow formulations of the previous chapter. Our exposition of these topics is extremely cursory but is just enough to carry the discussion to linear quadratic regulation, another application of convex optimization. The interested reader is referred to endnotes [7, 24–26] for in-depth coverage of power system dynamics.

The basic assumptions of Section 2.5 are especially pertinent here, because we are now closer to regimes in which they are not valid. In all of this section's modeling, we assume sinusoidal voltages and balanced three-phase operation, each of which is suspect at time scales smaller than a few cycles. Perhaps more significantly, we allow nodal frequencies to change dynamically but maintain the assumption that the lines are at the nominal steady-state frequency.[2] As such, while the control techniques we present are of general applicability, we take the power system modeling to be valid within a few percent of the nominal system frequency and at time scales no faster than a few cycles or a tenth of a second.

4.2.1 The swing equation

The swing equation describes the dynamic transfer of energy between generators and loads across a transmission network. Mathematically, it replaces the nodal power balance (e.g., (3.4)-(3.5)) with the dynamics at each node. At generators, this means the rotor torque balance and the terminal voltage. Load dynamics vary widely across load types.

The swing equation is highly nonlinear, so large-scale optimal control is out of the question with current techniques. Whereas nonlinear optimization can be easy if the objective and constraints are convex, most nonlinear control problems are extremely hard. From an optimization perspective, this is because a controlled differential equation of the form

$$\dot{x} = f(x, u)$$

is essentially an infinite family of equality constraints and is thus convex only if it is linear. This becomes even more apparent in discrete time, cf. (4.2) and (4.3). The reader is referred to endnotes [27–29] for further reading on nonlinear control.

If we apply the linearized approximation of Section 3.2.1, the swing equation becomes linear and thus amenable to a plethora of control techniques. Here we only present the basic linear quadratic regulator, which we return to in Section 6.2.2 for developing economic tools for regulation.

Generator modeling

We begin with the standard third order generator approximation. We must first introduce some parameters. Let T_i be the generator's direct axis transient short circuit time constant, and x_{di} and x'_{di} its direct axis synchronous and transient reactances. Let e_i be the constant internal field voltage and f_i be the field excitation. Finally, let J_i, D_i, and p_i^m be the rotor's inertia, damping coefficient, and mechanical power input. The third order generator model is then given by:

---

[2] Specifically, suppose that a line's resistance, inductance, and capacitance are denoted r, L, and C. Then its phasor impedance is $z = r + i\left(\omega L - \frac{1}{\omega C}\right)$, and its admittance is $y = g - ib = \left(r - i\left(\omega L - \frac{1}{\omega C}\right)\right) / \left(r^2 + \left(\omega L - \frac{1}{\omega C}\right)^2\right)$, where ω is the nominal system frequency in radians per second. This section keeps line admittances constant despite making ω a dynamic state variable at each node.

$$J_i \ddot{\theta}_i = -D_i \dot{\theta}_i + p_i^m - \sum_j g_{ij} |v_i|^2$$

$$-|v_i||v_j| \left(g_{ij} \cos\left(\theta_i - \theta_j\right) - b_{ij} \sin\left(\theta_i - \theta_j\right) \right) \quad (4.12)$$

$$\frac{d|v_i|}{dt} = \frac{2}{T_i} \left(\left((f_i + e_i) - |v_i|\right) - (x_{di} - x'_{di}) \sum_j b_{ij} |v_i| \right.$$

$$\left. -|v_j| \left(g_{ij} \sin\left(\theta_i - \theta_j\right) + b_{ij} \cos\left(\theta_i - \theta_j\right) \right) \right) \quad (4.13)$$

The control inputs in this model are typically p_i^m and f_i. Note that the right-hand side of the latter equation contains the expression for reactive power divided by $|v_i|$.

Often, the mechanical power input is modeled as the output of a speed governor, in which case p_i^M and the steam valve position, p_i^v, are themselves states with the approximately linear dynamics

$$\dot{p}_i^M = \frac{1}{t_c} \left(p_i^v - p_i^M \right)$$

$$\dot{p}_i^v = \frac{1}{t_g} \left(p_i^R - p_i^v - \frac{1}{\chi} \dot{\theta}_i \right)$$

In this case, control parameters can be the reference power, p_i^R, and the gain χ.

Load modeling
Whereas generators usually fall into one of a few categories and have precise physical models, there is a vast array of load types, which often are considered in aggregation. Loads are relatively slow compared to power system dynamics, and in many cases it is adequate to represent loads as static real and reactive power extractions, p_i and q_i. The result is dynamic equations such as those above at generation nodes coupled with the below algebraic power flow equations at load nodes:

$$p_i = \sum_j g_{ij} |v_i|^2 - |v_i||v_j| \left(g_{ij} \cos\left(\theta_i - \theta_j\right) - b_{ij} \sin\left(\theta_i - \theta_j\right) \right)$$

$$q_i = \sum_j b_{ij} |v_i|^2 - |v_i||v_j| \left(g_{ij} \sin\left(\theta_i - \theta_j\right) + b_{ij} \cos\left(\theta_i - \theta_j\right) \right)$$

While seemingly simpler, these *differential-algebraic* or *singular* systems are generally more complex to analyze than purely differential systems [30–32].

Algebraic equations can be avoided by modeling loads dynamically, cf. [7, 33], although the resulting system may still pose numerical difficulties due to multiple time scales. Since in many cases loads are inherently uncertain, it is common to parameterize linear dynamic models using empirical measurements.

The above models comprise the nonlinear swing equation and accurately capture the continuous dynamics and stability of AC power systems when operating close to the nominal system frequency. Instability can take many forms but is most often quantified in terms of frequency instability and voltage collapse. Frequency instability is when the voltage angle differences, $\theta_i - \theta_j$, become large, and $\dot{\theta}_i$ diverges from the nominal system frequency across the network. Voltage collapse is when $|v_i|$ everywhere sinks to zero. In actuality, frequency instability and voltage collapse are two sides of the same coin, which is to say that they represent two facets of dynamic instability within the same extremely complex system. Often, these instabilities are exacerbated by sequences of failures known as *cascading failure*. For example, the loss of a transmission line can cause a parallel line to carry too much power and trip out, in turn causing more lines and generators to disconnect. It is common to observe all of these phenomena in the few seconds leading up to a blackout.

Understanding the basic stability properties of the nonlinear swing equation has been the subject of much work, particularly via Lyapunov-based approaches [34–36]. A number of nonlinear control techniques have been applied as well [37, 38]. Additionally challenging to our current focus is that the nonlinear swing equation is not apparently amenable to the convex relaxations of Section 3.3; no relaxed coordinate system is known that isolates the voltage angle for use in the inertia and damping terms.

There is a full spectrum of simplifications between the above model and the linear swing equation corresponding to the linearized power flow, including:

- Neglect reactive power and generator voltage dynamics, (4.13), leaving only the inertial dynamics of generator rotors, (4.12).
- Set $g_{ij} = 0$ in (4.12).
- Set all voltages to one. The real power flow in a line is then approximated by $b_{ij} \sin (\theta_i - \theta_j)$. This model resembles the classical Kuramoto oscillator [39, 40].
- Linearize $\sin (\theta_i - \theta_j)$.

The final simplification yields the linear swing equation,

$$J_i \ddot{\theta}_i + D_i \dot{\theta}_i + \sum_j b_{ij} (\theta_i - \theta_j) = p_i^m,$$

where we set J_i and D_i equal to zero at load nodes. Due to the assumptions of linearized power flow, the linear swing equation only approximates frequency instability and does not describe voltage collapse. Likewise, the mechanical power inputs to the generators, p_i^m, are the only control variables.

Switching to matrix notation, we can straightforwardly eliminate any algebraic load equations using a technique known as *Kron reduction* [41, 42]. Let J and D be the appropriate diagonal inertia and damping matrices, and let the inductance matrix be defined as

$$B_{ij} = \begin{cases} -b_{ij} & \text{if } i \neq j \\ \sum_j b_{ij} & \text{if } i = j \end{cases}.$$

The matrix form of the linear swing equation is then given by

$$J\ddot{\theta} + D\dot{\theta} + B\theta = p^m. \tag{4.14}$$

Let $\mathcal{G}$ be the set of generators and $\mathcal{L}$ be the set of loads, and number the nodes such that

$$B = \begin{bmatrix} B_\mathcal{G} & B_{\mathcal{GL}} \\ B_{\mathcal{GL}}^T & B_\mathcal{L} \end{bmatrix} \quad \text{and} \quad \theta = \begin{bmatrix} \theta_\mathcal{G} \\ \theta_\mathcal{L} \end{bmatrix}.$$

The Kron-reduced version of B (and J and D) are then

$$B_\mathcal{K} = B_\mathcal{G} - B_{\mathcal{GL}} B_\mathcal{L}^{-1} B_{\mathcal{GL}}^T$$
$$J_\mathcal{K} = J_\mathcal{G}$$
$$D_\mathcal{K} = D_\mathcal{G},$$

which is well defined because $B_\mathcal{L}$ has full rank as long as a path exists from each load to at least one generator.[3] The following purely differential system is then dynamically equivalent to (4.14):

$$J_\mathcal{K} \ddot{\theta}_\mathcal{G} + D_\mathcal{K} \dot{\theta}_\mathcal{G} + B_\mathcal{K} \theta_\mathcal{G} = p_\mathcal{G}^m. \tag{4.15}$$

This is a *linear time-invariant system*, which is amenable to a vast array of powerful and scalable control techniques.

4.2.2   Linear quadratic regulation

Whereas most problems of analysis and control of large nonlinear dynamic systems are incredibly difficult analytically and computationally, many of the same problems are highly tractable for linear systems, to the extent that linear system theory is sometimes said to be "solved." Of course, there are many open problems in linear systems, particularly in regard to uncertainty, decentralization, and inequality constraints. And yet, the mathematical core of linear systems is extremely well understood, such that nearly all research topics are extensions to the basic framework.

Here we present the classical *linear quadratic regulator* [17, 43, 44], which is the solution to the following (informally stated) optimization problem:[4]

$$\underset{u}{\text{minimize}} \quad \frac{1}{2} \mathbb{E} \left[\int_0^T x^T Q x + u^T R u \; dt \right]$$
$$\text{subject to} \quad \dot{x} = Fx + Gu + w, \quad w \sim \mathcal{N}(0, W).$$

In this problem, x is the system state; for example, the vector of voltage angles and frequencies. Oftentimes, x is the deviation from the operating point about which the system was linearized. u is the control input; for example, an automatic generation control signal or reactive power injection. The Q matrix is positive semidefinite and associates quadratic costs with the states. The R matrix is positive definite and similarly associates quadratic costs with the control inputs. $\mathbb{E}$ denotes expectation over the disturbance w, which is Gaussian with zero mean and covariance W.

---

[3] Observe that the Kron-reduced version of B is just its Schur complement with respect to the load nodes, which we discussed in Section 2.2.2.

[4] Because the optimal controller does not depend on the starting and terminal points and we are primarily concerned with regulating disturbances, we omit the standard initial condition and terminal costs from our presentation.

While of a different form than the other optimization problems we have encountered, this is essentially an equality-constrained quadratic program and admits a simple optimal control policy. The continuous time Riccati equation is given by

$$\dot{P} + F^T P + PF + Q = PGR^{-1}G^T P, \quad P(T) = 0.$$

The optimal policy is then a linear function of the state given by

$$u = -R^{-1}G^T Px, \tag{4.16}$$

and the associated optimal objective by

$$\text{tr} \int_0^T PW \, dt.$$

While the disturbance covariance W determines the cost, it has no bearing on the optimal controller; this is known as the *certainty equivalence* property.

If we allow the time horizon T to approach infinity and consider the steady-state regulator, we obtain the continuous-time algebraic Riccati equation:

$$F^T P + PF + Q = PGR^{-1}G^T P, \quad P \succeq 0.$$

There are in fact a number of feasible solutions P, but only one is positive semidefinite, and this is the one corresponding to the optimal controller, also given by (4.16). In this case, the optimal *average* cost is given by

$$\text{tr } PW.$$

The linear quadratic regulator is the foundation of a number of more sophisticated techniques such as robust, hybrid, and constrained control, cf. [21, 44, 45]. The following example sketches the application of the linear quadratic regulator to frequency regulation.

Example 4.3 *Frequency regulation.* In this example, we write (4.15) as a first-order linear system so that the linear quadratic regulator can be applied to frequency regulation. The result will be a control policy for each generator's mechanical power input.

The linear quadratic regulator was first applied to frequency regulation some time ago in endnotes [46, 47] but was essentially unimplementable at the system scale due to limited communications. *Phasor measurement units*, also known as *synchrophasors*, now somewhat overcome this limitation. Using global positioning system satellite signals to *timestamp* measurements, they provide synchronized voltage and current phasor measurements approximately every cycle, which are processed and shared by *wide area measurement systems* [48, 49]. In other words, they provide the x in (4.16).

We designate voltage frequency by ω, so that the state is given by

$$x = \begin{bmatrix} \omega \\ \theta \end{bmatrix}.$$

The linearized, controlled swing equation is then parameterized by the matrices

$$F = \begin{bmatrix} -J_\mathcal{K}^{-1}D_\mathcal{K} & -J_\mathcal{K}^{-1}B_\mathcal{K} \\ \omega_0 I & 0 \end{bmatrix}$$

$$G = \begin{bmatrix} J_\mathcal{K}^{-1} \\ 0 \end{bmatrix}$$

$$W = \begin{bmatrix} W_\omega & 0 \\ 0 & 0 \end{bmatrix},$$

where ω_0 is the nominal system frequency in radians per second, e.g., 100π or 120π. Note that the control input and Gaussian disturbance only affect the torque balance in the first row. Also note that this system is marginally stable, which is problematic for some numerical implementations. This can be remedied by simply subtracting a very small positive number from each entry of the state transition matrix's main diagonal.

In the objective, each generator's voltage frequency deviation and control input are quadratically penalized as $q_i\omega_i^2$ and $r_i p_i^{m2}$, so that the objective matrices are

$$Q = \begin{bmatrix} \mathrm{diag}[q] & 0 \\ 0 & 0 \end{bmatrix}$$

$$R = \mathrm{diag}[r],$$

where $\mathrm{diag}[q]$ is a square matrix with the vector q on its main diagonal. The optimal policy for p^m is then given by (4.16).

4.3   Unit commitment

Thus far in this book, we have assumed that power may be drawn in continuous quantities from generation resources, or *units*. This is a reasonable and necessary approximation for power system operations on shorter time scales. On longer time scales, units' online and offline times must be scheduled in advance. Because many types of power plants require hours to days turn on or off, binary variables are required to model generator statuses, resulting in a fairly large mixed-integer programming problem known as unit commitment [1, 50, 51]. Unit commitment is essential to feasible power system operation. For this reason and its dramatic impact on power system efficiency and reliability, unit commitment is also a popular framework for assessing the influence of broader changes to power systems like renewable energy integration [5, 52]. Note that although power outputs are variables in unit commitment routines, here we are primarily interested in scheduling on- and offline times. Actual power outputs are then chosen after startup and shutdown decisions are made, using more detailed optimal power flow models and more recent information.

Here, we follow the unit commitment formulation of endnote [51]. Binary variables are used to represent whether a unit is on or off in each period and to enforce various logical conditions as in the disjunctive constraint in Example 2.6. With N units and a

Table 4.1 Unit commitment variables and parameters.

Variables	
ξ_i^t	Binary variable indicating unit on- and offline statuses
p_i^t	Real power output
$\overline{p}_i^t$	Maximum available real power output
CP_i^t	Real power production cost
CU_i^t	Startup cost
CD_i^t	Shutdown cost

Parameters	
T	Time horizon
d^t	Predicted demand
r^t	Predicted reserve requirement
$\hat{p}_i$	Maximum real power output
a_i, b_i, c_i	Quadratic production cost function coefficients of unit i
$\overline{CU}_i^k$	Startup cost after being offline for k periods
$\overline{CD}_i$	Shutdown cost
U_i	Minimum up time
U_i^0	Minimum initial up time
D_i	Minimum down time
D_i^0	Minimum initial down time
SU_i	Maximum startup ramp limit
SD_i	Maximum shutdown ramp limit
RU_i	Maximum ramp up limit
RD_i	Maximum ramp down limit

time horizon T, there are NT binary variables. There are also T power flow feasible sets, like in the multi-period optimal power flow of Section 4.1. For this reason and because the purpose of unit commitment is merely to determine which generators to turn on, it is common to model unit commitment in minimal detail. In particular, reactive power and voltage are almost never included, and power flow constraints are often condensed into a real power balance. We thus also presented unit commitment in minimal detail but indicate how to incorporate further details into the formulation. Because the set of unit commitment variables and parameters is somewhat large, they are collected in Table 4.1.

4.3.1 Objective

The unit commitment objective is to minimize three costs: power production, CP_i^t; startup, CU_i^t; and shutdown, CD_i^t – each summed through time over all units:

$$\text{maximize} \sum_t \sum_i CP_i^t + CU_i^t + CD_i^t. \tag{4.17}$$

The production cost is the standard quadratic cost curve discussed in Section 3.1:

$$CP_i^t = a_i \xi_i^t + b_i p_i^t + c_i \left(p_i^t\right)^2.$$

This sort of quadratic cost curve is often approximated with piecewise linear functions so that the resulting optimization is an MILP and because some markets require generators to submit step-wise cost curves. This is, however, inessential because MIQP has similar tractability to MILP, as discussed in Section 2.2.4, and so we omit this step. While more detailed cost curve modeling is viable (see Section 3.5), it would be somewhat disproportionate with the resolution at which the rest of the problem is modeled.

The startup cost for each unit, CU_i^t, is a continuous variable but can only take on a discrete set of values that are determined by when the unit is on- and offline. When $\xi_i^t = 1$, the parenthesized term in (4.18) is equal to one when k is the number of consecutive prior periods that the unit was offline and otherwise is an inactive constraint.

$$CU_i^t \geq \overline{CU}_i^k \left(\xi_i^t - \sum_{n=1}^{k} \xi_i^{t-n} \right), \quad k = 1, \ldots, |\overline{CU}_i| \tag{4.18}$$

$$CU_i^t \geq 0 \tag{4.19}$$

The underlying continuous startup cost is often nonconvex, for instance $\overline{CU}_i^k = 1 - e^t\big|_{t=k}$ is a simple approximation that is concave. This is, however, an inconsequential nonconvexity because it is absorbed by the binary constraints.

The shutdown cost is similarly discrete-valued and is inactive except when $\xi_i^{t-1} = 0$ and $\xi_i^t = 1$:

$$CD_i^t \geq \overline{CD}_i \left(\xi_i^{t-1} - \xi_i^t \right) \tag{4.20}$$

$$CD_i^t \geq 0. \tag{4.21}$$

Note that, unlike the startup cost, this does not depend on the unit's prior statuses.

4.3.2   Constraints

The defining feature of unit commitment is the set of NT binary constraints on the generators' on- and offline statuses:

$$\xi_i^t \in \{0, 1\}. \tag{4.22}$$

Until all large power sources can rapidly switch on and off, stay permanently online, or be partially on and partially off, these constraints will continue to complexify power system operation. While somewhat outlandish now, these may be realistic properties for power systems with only renewable sources like wind and solar.

It is useful to distinguish between the power a unit can actually provide and its maximum power capacity. We now regard $\overline{p}_i^t$, the maximum *available* power output, as a variable quantity and denote the maximum *possible* power output by $\hat{p}_i$. The minimum online power output, $\underline{p}_i$, remains a constant parameter. These quantities are related to the actual power output, p_i^t, by the following two constraints:

$$\underline{p}_i \xi_i^t \leq p_i^t \leq \overline{p}_i^t \tag{4.23}$$

$$0 \leq \overline{p}_i^t \leq \hat{p}_i \xi_i^t \tag{4.24}$$

The reason for distinguishing the maximum available power is *reserves*, power capacity kept online to accommodate failures and forecast errors.

In the simplest possible formulation, the total power produced in each period balances the predicted demand, and the total available power is at least the forecasted demand plus the prescribed reserve capacity requirement:

$$\sum_i p_i^t = d^t \tag{4.25}$$

$$\sum_i \bar{p}_i^t \geq d^t + r^t. \tag{4.26}$$

To include voltage and transmission modeling, one could decompose d^t and $d^t + r^t$ into predicted demand and reserve requirements at each node and then replace each constraint in the above pair with a feasible set of power flows from Chapter 3. Clearly, there is some flexibility in how much detail can be allowed into the model, e.g., the linearized power flow would lead to a far more tractable transmission-constrained problem than incorporating voltage and transmission constraints via an SOC or SD relaxation.

Observe that the reserve requirement, (4.26), incorporates load uncertainty by requiring the power capacity to exceed the worst-case load under-prediction, r^t. This is a very simple instance of robust optimization, which we discuss briefly in Section 7.1.2.

The minimum up and down times specify how long a generator must remain on- and offline once turned on and off, respectively. The following two constraints determine how long a generator must initially be on- or offline due to its activity prior to the optimization time interval; note that only one of the two is included for each unit.

$$\xi_i^t = 1, \quad t = 1, \ldots, U_i^0 \quad \text{or} \quad \xi_i^t = 0, \quad t = 1, \ldots, D_i^0. \tag{4.27}$$

For example, if unit i's minimum up time is $U_i = 6$ and it was online for two periods prior to $t = 1$, $U_i^0 = 4$.

The following two constraints ensure that if a unit goes online at period t, it stays online for at least the minimum up time, U_i. The latter handles periods less than $U_i - 1$ from the final period T.

$$\sum_{k=t}^{t+U_i-1} \xi_i^k \geq U_i \left(\xi_i^t - \xi_i^{t-1} \right), \quad t = U_i^0 + 1, \ldots, T - U_i + 1 \tag{4.28}$$

$$\sum_{k=t}^{T} \xi_i^k \geq (T - t + 1) \left(\xi_i^t - \xi_i^{t-1} \right), \quad t = T - U_i + 2, \ldots, T - 1 \tag{4.29}$$

Similarly, the next two constraints ensure that when a unit goes offline, it stays offline for the minimum down time, D_i.

$$\sum_{k=t}^{t+D_i-1} 1 - \xi_i^k \geq D_i \left(\xi_i^{t-1} - \xi_i^t \right), \quad t = D_i^0 + 1, \ldots, T - D_i + 1 \tag{4.30}$$

$$\sum_{k=t}^{T} 1 - \xi_i^k \geq (T - t + 1) \left(\xi_i^{t-1} - \xi_i^t \right), \quad t = T - D_i + 2, \ldots, T - 1 \tag{4.31}$$

The final constraints encode each unit's ramping capabilities, similar to those discussed in Section 4.1.1. The first ramp constraint says that the maximum available power, $\bar{p}_i^t$, can be no more than that of the previous period plus the maximum amount it can ramp up if it was online, or the maximum startup ramp if it was previously offline. When offline, the final term deactivates the constraint.

$$\bar{p}_i^t \leq p_i^{t-1} + RU_i \xi_i^{t-1} + SU_i \left(\xi_i^t - \xi_i^{t-1} \right) + \hat{p}_i \left(1 - \xi_i^t \right) \tag{4.32}$$

The down ramping analog has signs reversed but is otherwise similar:

$$p_i^t \geq p_i^{t-1} - RD_i \xi_i^t - SD_i \left(\xi_i^{t-1} - \xi_i^t \right) - \hat{p}_i \left(1 - \xi_i^{t-1} \right) \tag{4.33}$$

Finally, the maximum available power is lower than the maximum shutdown ramp limit if it is going offline in the next period:

$$\bar{p}_i^{t-1} \leq SD_i \left(\xi_i^{t-1} - \xi_i^t \right) + \hat{p}_i \xi_i^t \tag{4.34}$$

Note that (4.33) is not made trivial by (4.23) when the unit is shutting down. Rather, it ensures that p_i^{t-1} is small enough that it is able to shutdown at period t.

The above constraints are composed in the following feasible set.

FEASIBLE SET 4.2 (MIL unit commitment) (4.18)-(4.34)

Observe that unit commitment is essentially the multi-period optimal power flow of Section 4.1 with additional constraints and binary variables (which render unit commitment far less computationally tractable than multi-period optimal power flow). Our modeling approach grants us some liberty to pick and choose different details; for example, we might differentiate between fast and slow reserves, r_f^t and r_s^t, and group the generators according to which type they can provide. Transmission constrained models are obtained by replacing the power balances, (4.25) and (4.26), with a feasible set from Chapter 3, as illustrated in Example 4.4.

Example 4.4 *Unit commitment with the linearized power flow.* To incorporate transmission constraints into unit commitment models, we must associate units with individual nodes. Let $\mathcal{N}_i$ be the set of units located at node i, and denote the demand at node i by d_i^t. The transmission constrained model is obtained by replacing (4.25) with the nodal real power balance

$$\sum_{k \in \mathcal{N}_i} p_k^t = d_i^t + \sum_j p_{ij}^t,$$

where the power flow from node i to j at time t, p_{ij}^t, is constrained by Feasible Set 3.3 at each period. This amounts to the additional constraints

$$p_{ij}^t = b_{ij} \left(\theta_i^t - \theta_j^t \right)$$

$$\left| p_{ij}^t \right| \leq \bar{s}_{ij}.$$

Because reserves need not be associated with any particular node, (4.26) can be replaced with

$$\sum_i -d_i^t + \sum_{k \in \mathcal{N}_i} \bar{p}_k^t \geq r^t.$$

Observe that although the linearized power flow inevitably makes unit commitment harder to solve than power balance formulations, it does not take it out of MILP. Hence, both types of models are amenable to most of the same basic algorithms.

Example 4.4 can be easily adapted to a relaxed power flow such as Feasible Set 3.8 by including reactive power versions of the appropriate constraints from Feasible Set 4.2. As usual, however, the added complexity of this level of realism may outweigh its benefit when making decisions as coarse as unit startups and shutdowns.

4.4   Reconfiguration

Distribution systems are usually operated radially, but multiple radial topologies may be attainable by opening and closing different combinations of protective switches, as shown in Figure 4.3. Each switch can only be open or closed and must thus be modeled as a binary variable, leading to an exponentially large number of switch configurations. Choosing the best one is known as *reconfiguration*, as introduced in endnote [53]. Coincidentally, the modern version of reconfiguration and the initial radial version of the branch flow model of Section 3.3.3 were presented together in endnote [54]. Since then, reconfiguration has become one of the most written about research problems in power systems. Here, we present reconfiguration using SOC relaxations, cf. [55, 56]. Because reconfiguration always results in a radial network, the SOC relaxations are ideal fits. We also consider the more recent transmission switching problem [57, 58], which, as will be seen, is obtained by simply taking a subset of the constraints that make up reconfiguration.

Define the following four sets:

- The node set, $\mathcal{N}$.
- The set of root nodes, $\mathcal{N}^r \subseteq \mathcal{N}$. These are nodes through which power enters the network, e.g., substations or local power sources.
- The undirected line set, $\mathcal{L}$. There is one element in $\mathcal{L}$ for each line.
- The set of lines with switches, $\mathcal{L}^s \subseteq \mathcal{L}$.

We now make use of some graph theory from Section 2.3.2. Together, the set of nodes and lines constitute an undirected graph, which we denote $(\mathcal{N}, \mathcal{L})$. A subset of $(\mathcal{N}, \mathcal{L})$ is a subgraph, for example the graph of switches, $(\mathcal{N}, \mathcal{L}^s)$. We assume that all nodes in

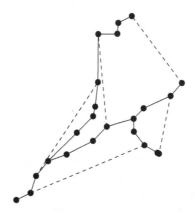

Fig. 4.3   A distribution system with multiple radial configurations. Solid lines have closed switches or no switches, and dashed lines have open switches.

$\mathcal{N}/\mathcal{N}^r$ consume real and reactive power.[5] This assumption is sufficient to ensure that all real and reactive power flows downstream (away from the root node) in any radial configuration. The orientation of the power flow on a line thus defines its direction, inducing a rooted, directed radial graph or an *arborescence* on $(\mathcal{N}, \mathcal{L})$.

4.4.1   Radiality constraints

To formulate reconfiguration as an optimization, we must first construct a set of constraints ensuring that any feasible configuration is radial. Associate with each undirected line $ij \in \mathcal{L}$ two continuous variables, z_{ij} and z_{ji}. We will have $z_{ij} = 1$ if power flows from i to j, and $z_{ij} = 0$ otherwise. If $ij \in \mathcal{L}^s$, also associate with it a single binary variable, ξ_{ij}, representing if the switch is open or closed. Note that since the lines themselves are nondirectional, there are the same number of binary variables as switches. The following set of constraints ensures that any feasible network is radial.

FEASIBLE SET 4.3 (MIL radiality constraints)

$$z_{ij} \geq 0 \tag{4.35}$$

$$z_{ij} = 0, \quad j \in \mathcal{N}^r \tag{4.36}$$

$$z_{ij} + z_{ji} = 1, \quad ij \in \mathcal{L} \backslash \mathcal{L}^s \tag{4.37}$$

$$z_{ij} + z_{ji} = \xi_{ij}, \quad ij \in \mathcal{L}^s \tag{4.38}$$

$$\sum_j z_{ji} = 1, \quad i \in \mathcal{N} \backslash \mathcal{N}^r \tag{4.39}$$

$$\xi_{ij} \in \{0, 1\}, \quad ij \in \mathcal{L}^s \tag{4.40}$$

---

[5] While historically very reasonable, this assumption is now at odds with current trends toward distributed generation, energy storage, and demand response. Nevertheless, the problem remains of great practical significance, particularly on longer time scales over which most loads consume power on average.

Define the set of lines without switches or with closed switches, $\mathcal{L}^{\xi} = \{ij \in \mathcal{L}^s : \xi_{ij} = 1\}$ $\cup (\mathcal{L} \backslash \mathcal{L}^s)$. A switching configuration ξ induces the subgraph $(\mathcal{N}, \mathcal{L}^{\xi}) \subseteq (\mathcal{N}, \mathcal{L})$. We now show that if ξ is feasible, $(\mathcal{N}, \mathcal{L}^{\xi})$ is radial.

THEOREM 4.1 *Assume that a path exists from every non-root node to at least one root node in the graph* $(\mathcal{N}, \mathcal{L}^{\xi})$. *Then if ξ is feasible for Feasible Set 4.3,* $(\mathcal{N}, \mathcal{L}^{\xi})$ *is radial and no two root nodes are connected.*

Proof Assume ξ is feasible. Consider an arbitrary path through $(\mathcal{N}, \mathcal{L}^{\xi})$ beginning at a root node, which we label node zero. We proceed inductively along the path. By (4.36) and (4.37), $z_{i0} = 0$ and $z_{0i} = 1$ for any node i connected to 0. Now suppose that line jk is on the path, j is between the root and k, and $z_{jk} = 1$. By (4.39), $z_{lk} = 0$ and, by (4.37) and (4.38), $z_{kl} = 1$ for all $l \neq j$, $lj \in \mathcal{L}^{\xi}$. Therefore, for any node j on the path, $z_{ij} = 1$ if i is between j and a root node and $z_{ij} = 0$ otherwise. Because we have considered an arbitrary path and assumed that each node is connected to a root node, $z_{ij} = 1$ if i is between a root node and j, and $z_{ij} = 0$ otherwise for each undirected line $ij \in \mathcal{L}^{\xi}$.

Suppose for the sake of contradiction that a loop exists in $\mathcal{L}^{\xi}$. By our initial assumption, it must be connected to a root node. Suppose that the path from the root node enters the loop at node j and returns there via the loop (the root node is j if it is in the loop). Then, by the above argument, $z_{ij} = 1$ for two different nodes i, violating (4.39). Therefore, no loops exist, and the graph is radial.

Now suppose that a path connects two root nodes. Then there exists a node j between them for which $z_{ij} = 1$ for two different nodes i, also violating (4.39). Therefore, root nodes cannot be connected to each other. ☐

Observe that the integer constraint has no effect on the network's radiality. Rather, the first five constraints enforce that the graph induced by ξ, $(\mathcal{N}, \mathcal{L}^{\xi})$, is radial. Intuitively, z represents the direction power can flow through each line. z thus describes a collection of arborescences on the undirected graph $(\mathcal{N}, \mathcal{L}^{\xi})$. The assumption that a path exists between every non-root node and a root node merely prohibits the existence of unpowered islands. If the root nodes are the only power sources, the assumption becomes unnecessary upon addition of the below power flow constraints.

4.4.2   Power flow and objectives

The feasible set for reconfiguration is obtained by combining the above radiality constraints with any of the power flow feasible sets from Section 3.3. Because any feasible configuration must be radial, the associated power flow is likely to be exactly described by the SOC relaxation due to the exactness result of Section 3.3.1, nullifying any gains from using an SDP relaxation or non-relaxed power flow.

One adjustment must be made in fusing Feasible Set 4.3 with a power flow: the power flow constraints on a line with a switch should only be enforced if the switch is closed. This is straightforwardly accomplished using disjunctive constraints, as described in Example 2.6. Let $M > 0$ be a sufficiently large number, and let P_{ij} and Q_{ij} be the power flow on line ij were its switch closed. For example, if the voltage-polar coordinate form of Feasible Set 3.10 was used, we would have

$$P_{ij} = g_{ij}W_{ii} - g_{ij}W_{ij} + b_{ij}W_{i+n,j} \tag{4.41}$$

$$Q_{ij} = b_{ij}W_{ii} - g_{ij}W_{i+n,j} - b_{ij}W_{ij} \tag{4.42}$$

$$W_{ij} = W_{i+n,j+n} \tag{4.43}$$

$$W_{i,j+n} = -W_{i+n,j} \tag{4.44}$$

$$W_{ij}^2 + W_{i+n,j}^2 \le W_{ii}W_{jj}. \tag{4.45}$$

The following disjunctive constraints equate P_{ij} and Q_{ij} with the actual power flows, p_{ij} and q_{ij}, only when $\xi_{ij} = 1$:

$$|p_{ij}| \le M\xi_{ij} \tag{4.46}$$

$$|q_{ij}| \le M\xi_{ij} \tag{4.47}$$

$$|p_{ij} - P_{ij}| \le M(1 - \xi_{ij}) \tag{4.48}$$

$$|q_{ij} - Q_{ij}| \le M(1 - \xi_{ij}). \tag{4.49}$$

Observe that since ξ_{ij} is nondirectional, the above constraints either block all flows or allow flow in either direction on line ij. In the case that $\xi_{ij} = 0$, (4.41) and (4.42) do not constrain the associated entries of W because P_{ij} and Q_{ij} are free.

The reconfiguration feasible set is the following.

FEASIBLE SET 4.4 (MISOC reconfiguration) Feasible Sets 3.10 [SOC power flow relaxation] and 4.3 [Radiality], (4.46)-(4.49).

There are a handful of reconfiguration objectives, the most common of which is resistive power losses:

$$\sum_{ij} p_{ij} + p_{ji}.$$

Note that because all flow enters through root nodes, the objective

$$\sum_{i \in \mathcal{N}^r} p_i$$

also suffices for resistive loss reduction. Load balancing is another popular objective that attempts to level the ratios of the lines' apparent power flows and capacities to reduce maintenance costs and the risk of overcapacity failures. It is given by the linear objective and additional convex quadratic constraints

$$\text{minimize } t \quad \text{subject to} \quad p_{ij}^2 + q_{ij}^2 \le \bar{s}_{ij}^2 t.$$

Finally, switching operations themselves may incur maintenance costs and can be included as a secondary objective. Denote the initial switch configuration by ξ_{ij}^0 and the cost of toggling the switch on line ij by c_{ij}. The total cost of switching is simply the total cost of all switching operations,

$$\sum_{ij \in \mathcal{L}^s} c_{ij} \left| \xi_{ij} - \xi_{ij}^0 \right|.$$

Of course, any combination of the above or other (convex) objectives may be used.

As presented, reconfiguration is very naturally an MISOCP. As discussed in Section 2.2.5, MISOCP is rapidly developing but does not yet enjoy the scalability of MILP or MIQP. There may thus be substantial computational benefit in applying the linear network flow relaxation of Section 3.2.3, as shown in Example 4.5.

Example 4.5 *Loss reduction via mixed-integer quadratic reconfiguration.* In this example, we formulate reconfiguration using two-commodity network flow (see Section 3.2.3). Recall that in radial networks, network flow and the linearized power flow of Section 3.2.1 are identical. Thus, due to the radiality constraints, the linearized power flow should never be used for reconfiguration because network flow is mathematically equivalent and computationally simpler.

Because network flow is lossless, we must approximate the loss minimization objective. Assuming voltage magnitudes are approximately one per unit, resistive losses are given by

$$r_{ij} |I_{ij}|^2 = r_{ij} \frac{p_{ij}^2 + q_{ij}^2}{|v_i|^2}$$
$$\approx r_{ij} \left(p_{ij}^2 + q_{ij}^2 \right).$$

The full optimization is then given below.

$$\underset{\xi,z,p,q}{\text{minimize}} \quad \sum_{ij} r_{ij} \left(p_{ij}^2 + q_{ij}^2 \right)$$

$$\text{subject to} \quad \text{Feasible Set 4.3} \quad [\text{Radiality}]$$
$$|p_{ij}| \leq M\xi_{ij}$$
$$|q_{ij}| \leq M\xi_{ij}$$
$$p_{ij} + p_{ji} = 0$$
$$q_{ij} + q_{ji} = 0$$
$$\sum_j p_{ij} = p_i$$
$$\sum_j q_{ij} = q_i$$
$$\underline{p_i} \leq p_i \leq \overline{p_i}$$
$$\underline{q_i} \leq q_i \leq \overline{q_i}.$$

Notice that, because network flow enforces no physics save flow conservation, it is not necessary to include (4.48) and (4.49) in this model.

4.4.3　Transmission switching

Braess's paradox is a classical counterintuitive phenomenon seen in transportation networks where the addition of new routes can actually increase traffic congestion and vice versa for route removal [59]. Meshed power networks can exhibit a similar behavior in which opening a line's switch can actually decrease the cost of generation, cf. [60]. This observation is mechanized by a cost-saving procedure known as transmission switching [57, 58].

Mathematically, the problem is identical to distribution system reconfiguration without the radially constraint. The transmission switching feasible set is given below.

FEASIBLE SET 4.5 (Mixed-integer transmission switching) A power flow feasible set from Chapter 3, (4.46)-(4.49)

Unlike with reconfiguration, the power flow relaxations of Section 3.3 may not be exact here because the network need not be radial. This is, however, a minor point because most formulations begin with the linearized approximation given by Feasible Set 3.2.1, which is cruder than the relaxations. Note that, while network flow is suitable for reconfiguration due to its equivalence with the linearized approximation in radial networks, this is not the case here, and the latter is a better choice of linear approximation. We illustrate transmission switching with this model in Example 4.6.

Example 4.6　*Mixed-integer quadratic transmission switching.* Suppose that each generator's cost curve, $f_i(p_i)$, is convex. Transmission switching with the linearized power flow (Feasible Set 3.2.1) is given by:

$$\underset{\xi,p,\theta}{\text{minimize}} \quad \sum_i f_i(p_i)$$

$$\text{subject to} \quad \sum_j p_{ij} = p_i$$

$$\underline{p}_i \leq p_i \leq \overline{p}_i$$

$$|p_{ij}| \leq \overline{s}_{ij}\xi_{ij}$$

$$|p_{ij} - b_{ij}(\theta_i - \theta_j)| \leq M(1 - \xi_{ij})$$

$$\xi_{ij} \in \{0, 1\}.$$

Here we have consolidated a number of constraints from Feasible Set 4.5, namely the power flow limit $|p_{ij}| \leq \overline{s}_{ij}$ with (4.46) and $p_{ij} = b_{ij}(\theta_i - \theta_j)$ with (4.48). Note that the second-to-last constraint is almost the same as the disjunctive constraint used in transmission planning with the linearized model, (5.22). Indeed, transmission switching is mathematically identical to transmission planning in Section 5.2 when the decision to build each line is binary.

4.5   Summary

The purpose of this chapter has been twofold:

- To express certain prominent power system optimization problems in the language of convex optimization and
- To demonstrate the application of techniques and formulations from Chapters 2 and 3.

The result is a hodgepodge of problems loosely connected by the fact that most build on the same power flow constraints. We summarize a few salient observations below.

When radiality is a feature like in distribution systems or reconfiguration problems, the exactness result of Theorem 3.1 ensures the accuracy of the SOC power flow relaxations and guarantees that there are no further gains from using the generally more accurate SD relaxation.

Integer constraints are encountered in a number of contexts and often necessitate accepting the reduced accuracy of linear power flow models in exchange for computational tractability. However, as discussed in Section 2.2.5, algorithms for MISOCP and MISDP will be practical someday, at which time the relaxations will be unambiguously better choices than linear power flow approximations.

We reflect on this chapter by viewing its content as a rough template with which to formulate more nuanced models in terms of convex optimization. We explored this theme to a minor extent when we combined storage and ramp constraints in a multi-period optimal power flow in Example 4.2 and expand on it by developing more joint models in this chapter's exercises.

Problems

4.1   Construct a multi-period optimal power flow that uses the linearized power flow (Feasible Set 3.3) at certain time periods and the SOC relaxation (Feasible Set 3.10) at others. In what scenarios might such a formulation be useful?

4.2   Construct a heuristic version of the inventory control policy in Section 4.1.2 that incorporates energy storage injection and extraction losses as modeled in Feasible Set 4.1.

4.3   Augment Example 4.3 to include energy storage in frequency regulation. Express the constituent matrices of a linear quadratic regulator that has "softened" energy and power limits in its objective.

4.4   Reformulate transmission constrained unit commitment (Example 4.4) with a relaxed power flow such as Feasible Set 3.8.

4.5   Formulate reconfiguration with the objective of minimizing the maximum (per unit) voltage magnitude's deviation from one.

4.6 Consider the following heuristic for reconfiguration with the loss minimization objective. Let $\mathcal{L}^r$ be the graph with edge weights equal to r_{ij} induced by closing all switches, and suppose that there is only one root node, which is labeled zero. Let the distance between two nodes be defined as the sum of the edge weights on the path. The *shortest-path tree* is the radial subgraph for which the path between 0 and any other node is the shortest path in $(\mathcal{N}, \mathcal{L}^r)$. Construct an example for which the switching configuration corresponding to the shortest-path tree is suboptimal for the mixed-integer linear formulation in Example 4.5 with identical line resistances, i.e., $r_{ij} = r$ for all ij.

4.7 Give conditions for which the shortest-path tree heuristic in the previous exercise provably produces the optimal solution. Are the conditions realistic?

4.8 Consider a network with infinite line capacities, $x_{ij}/r_{ij} = \alpha$ for all lines ij, and switches on a subset of the lines. Either prove or find a counterexample to the claim that closing all switches minimizes total resistive losses.

4.9 Formulate multi-period optimal power flow with transmission switching.

4.10 Formulate multi-period optimal power flow with reconfiguration. Devise an objective that penalizes the total number of switching operations over all time periods.

References

[1] A. J. Wood and B. F. Wollenberg, *Power Generation, Operation, and Control*, 3rd ed. Wiley, 2013.

[2] P. Varaiya, F. Wu, and J. Bialek, "Smart operation of smart grid: Risk-limiting dispatch," *Proceedings of the IEEE*, vol. 99, no. 1, pp. 40–57, 2011.

[3] R. Rajagopal, E. Bitar, P. Varaiya, and F. Wu, "Risk-limiting dispatch for integrating renewable power," *International Journal of Electrical Power & Energy Systems*, vol. 44, no. 1, pp. 615–628, 2013.

[4] J. Warrington, P. Goulart, S. Mariethoz, and M. Morari, "Policy-based reserves for power systems," *IEEE Transactions on Power Systems*, vol. 28, no. 4, pp. 4427–4437, 2013.

[5] B. Ummels, M. Gibescu, E. Pelgrum, W. Kling, and A. Brand, "Impacts of wind power on thermal generation unit commitment and dispatch," *IEEE Transactions on Energy Conversion*, vol. 22, no. 1, pp. 44–51, 2007.

[6] B. Lu and M. Shahidehpour, "Short-term scheduling of combined cycle units," *IEEE Transactions on Power Systems*, vol. 19, no. 3, pp. 1616–1625, 2004.

[7] P. Kundur, *Power System Stability and Control*. McGraw-Hill Professional, 1994.

[8] J. Barton and D. Infield, "Energy storage and its use with intermittent renewable energy," *IEEE Transactions on Energy Conversion*, vol. 19, no. 2, pp. 441–448, June 2004.

[9] H. Ibrahim, A. Ilinca, and J. Perron, "Energy storage systems - characteristics and comparisons," *Renewable and Sustainable Energy Reviews*, vol. 12, no. 5, pp. 1221–1250, 2008.

[10] J. Tester, E. Drake, M. Driscoll, M. Golay, and W. Peters, *Sustainable Energy: Choosing Among Options*. MIT Press, 2012.

[11] S. Borenstein, M. Jaske, and A. Rosenfeld, "Dynamic pricing, advanced metering, and demand response in electricity markets," UC Berkeley: Center for the Study of Energy Markets, Tech. Rep., 2002.

[12] D. Callaway and I. A. Hiskens, "Achieving controllability of electric loads," *Proceedings of the IEEE*, vol. 99, no. 1, pp. 184–199, Jan. 2011.

[13] J. Mathieu, S. Koch, and D. Callaway, "State estimation and control of electric loads to manage real-time energy imbalance," *IEEE Transactions on Power Systems*, vol. 28, no. 1, pp. 430–440, 2013.

[14] A. Nayyar, J. A. Taylor, A. Subramanian, D. S. Callaway, and K. Poolla, "Aggregate flexibility of collections of loads," in *IEEE 52nd Annual Conference on Decision and Control (CDC)*, Dec. 2013, pp. 5600–5607, invited.

[15] R. Erickson and D. Maksimovic, *Fundamentals of Power Electronics*. Springer, 2001.

[16] J. Carrasco, L. Franquelo, J. Bialasiewicz, E. Galvan, R. Guisado, M. Prats, J. Leon, and N. Moreno-Alfonso, "Power-electronic systems for the grid integration of renewable energy sources: A survey," *IEEE Transactions on Industrial Electronics*, vol. 53, no. 4, pp. 1002–1016, June 2006.

[17] D. P. Bertsekas, *Dynamic Programming and Optimal Control, Two Volume Set*. Athena Scientific, 2005.

[18] L. Snyder and Z.-J. Shen, *Fundamentals of Supply Chain Theory*. John Wiley & Sons, 2011.

[19] J. A. Taylor, D. S. Callaway, and K. Poolla, "Competitive energy storage in the presence of renewables," *IEEE Transactions on Power Systems*, vol. 28, no. 2, pp. 985–996, 2013.

[20] E. F. Camacho, C. Bordons, E. F. Camacho, and C. Bordons, *Model Predictive Control*, 2nd ed. Springer, 2013.

[21] F. Borrelli, A. Bemporad, and M. Morari, *Predictive Control for Linear and Hybrid Systems*. Cambridge University Press, Forthcoming.

[22] P. Kundur, J. Paserba, V. Ajjarapu, G. Andersson, A. Bose, C. Canizares, N. Hatziargyriou, D. Hill, A. Stankovic, C. Taylor, T. Van Cutsem, and V. Vittal, "Definition and classification of power system stability IEEE/CIGRE joint task force on stability terms and definitions," *IEEE Transactions on Power Systems*, vol. 19, no. 3, pp. 1387–1401, Aug. 2004.

[23] G. Andersson, P. Donalek, R. Farmer, N. Hatziargyriou, I. Kamwa, P. Kundur, N. Martins, J. Paserba, P. Pourbeik, J. Sanchez-Gasca, R. Schulz, A. Stankovic, C. Taylor, and V. Vittal, "Causes of the 2003 major grid blackouts in North America and Europe, and recommended means to improve system dynamic performance," *IEEE Transactions on Power Systems*, vol. 20, no. 4, pp. 1922–1928, Nov. 2005.

[24] P. W. Sauer and M. A. Pai, *Power System Dynamics and Stability*. Stipes Publishing Co., 1998.

[25] J. Machowski, J. Bialek, and J. Bumby, *Power System Dynamics: Stability and Control*, 2nd ed. Wiley-IEEE Press, 2002.

[26] P. M. Anderson and A. A. Fouad, *Power System Control and Stability*, 2nd ed. John Wiley & Sons, 2008.

[27] J.-J. E. Slotine and W. Li, *Applied Nonlinear Control*. Prentice Hall, 1991.

[28] A. Isidori, *Nonlinear Control Systems*, 3rd ed. Springer, 1999.

[29] H. K. Khalil, *Nonlinear Systems*, 3rd ed. Prentice Hall, 2002.

[30] S. L. V. Campbell and S. Campbell, *Singular Systems of Differential Equations*. Pitman San Francisco, 1980.

[31] K. E. Brenan, S. L. Campbell, and L. R. Petzold, *Numerical Solution of Initial-value Problems in Differential-Algebraic Equations*. Society for Industrial and Applied Mathematics, 1987, vol. 14.

[32] L. Dai, *Singular Control Systems*. Springer, 1989.

[33] H. Renmu, M. Jin, and D. Hill, "Composite load modeling via measurement approach," *IEEE Transactions on Power Systems*, vol. 21, no. 2, pp. 663–672, 2006.

[34] A. Bergen and D. Hill, "A structure preserving model for power system stability analysis," *IEEE Transactions on Power Apparatus and Systems*, vol. 100, no. 1, pp. 25–35, Jan. 1981.

[35] M. Pai, *Energy Function Analysis for Power System Stability*. Springer, 1989.

[36] H.-D. Chiang, *Direct Methods for Stability Analysis of Electric Power Systems: Theoretical Foundation, BCU Methodologies, and Applications*. Wiley, 2010.

[37] J. Chapman, M. Ilic, C. King, L. Eng, and H. Kaufman, "Stabilizing a multimachine power system via decentralized feedback linearizing excitation control," *IEEE Transactions on Power Systems*, vol. 8, no. 3, pp. 830–839, Aug. 1993.

[38] Y. Wang, D. Hill, R. Middleton, and L. Gao, "Transient stability enhancement and voltage regulation of power systems," *IEEE Transactions on Power Systems*, vol. 8, no. 2, pp. 620–627, May 1993.

[39] Y. Kuramoto, *Chemical Oscillations, Waves, and Turbulence*. Dover Publications, 2003.

[40] F. Dorfler and F. Bullo, "Synchronization and transient stability in power networks and nonuniform Kuramoto oscillators," *SIAM Journal on Control and Optimization*, vol. 50, no. 3, pp. 1616–1642, 2012.

[41] G. Kron, *Tensor Analysis of Networks*. J. Wiley & Sons, 1939.

[42] F. Dorfler and F. Bullo, "Kron reduction of graphs with applications to electrical networks," *IEEE Transactions on Circuits and Systems I: Regular Papers*, vol. 60, no. 1, pp. 150–163, 2013.

[43] F. Lewis and V. Syrmos, *Optimal Control*, ser. A Wiley-Interscience publication. John Wiley & Sons, 1995.

[44] K. Zhou, J. C. Doyle, and K. Glover, *Robust and Optimal Control*. Prentice Hall, 1996.

[45] S. Boyd, L. El Ghaoui, E. Feron, and V. Balakrishnan, *Linear Matrix Inequalities in System and Control Theory*, ser. Studies in Applied Mathematics. Philadelphia: SIAM, 1994, vol. 15.

[46] O. Elgerd and C. Fosha, "Optimum megawatt-frequency control of multiarea electric energy systems," *IEEE Transactions on Power Apparatus and Systems*, vol. 89, no. 4, pp. 556–563, Apr. 1970.

[47] C. Fosha and O. Elgerd, "The megawatt-frequency control problem: A new approach via optimal control theory," *IEEE Transactions on Power Apparatus and Systems*, vol. 89, no. 4, pp. 563–577, Apr. 1970.

[48] A. Phadke, "Synchronized phasor measurements in power systems," *IEEE Computer Applications in Power*, vol. 6, no. 2, pp. 10–15, Apr. 1993.

[49] A. G. Phadke and J. S. Thorp, *Synchronized Phasor Measurements and Their Applications*. Springer, 2008.

[50] N. Padhy, "Unit commitment–a bibliographical survey," *IEEE Transactions on Power Systems*, vol. 19, no. 2, pp. 1196–1205, 2004.

[51] M. Carrion and J. Arroyo, "A computationally efficient mixed-integer linear formulation for the thermal unit commitment problem," *IEEE Transactions on Power Systems*, vol. 21, no. 3, pp. 1371–1378, Aug. 2006.

[52] A. Papavasiliou, S. Oren, and R. O'Neill, "Reserve requirements for wind power integration: A scenario-based stochastic programming framework," *IEEE Transactions on Power Systems*, vol. 26, no. 4, pp. 2197–2206, 2011.

[53] A. Merlin and H. Back, "Search for a minimal-loss operating spanning tree configuration in an urban power distribution system," in *Proc. of the Fifth Power System Conference (PSCC), Cambridge*, 1975, pp. 1–18.

[54] M. Baran and F. Wu, "Network reconfiguration in distribution systems for loss reduction and load balancing," *IEEE Transactions on Power Delivery*, vol. 4, no. 2, pp. 1401–1407, Apr. 1989.

[55] J. A. Taylor and F. S. Hover, "Convex models of distribution system reconfiguration," *IEEE Transactions on Power Systems*, vol. 27, no. 3, pp. 1407–1413, Aug. 2012.

[56] R. Jabr, R. Singh, and B. Pal, "Minimum loss network reconfiguration using mixed-integer convex programming," *IEEE Transactions on Power Systems*, vol. 27, no. 2, pp. 1106–1115, May 2012.

[57] S. Blumsack, *Network Topologies and Transmission Investment Under Electric-Industry Restructuring*. Carnegie Mellon University, 2006.

[58] E. Fisher, R. O'Neill, and M. Ferris, "Optimal transmission switching," *IEEE Transactions on Power Systems*, vol. 23, no. 3, pp. 1346–1355, 2008.

[59] D. Braess, A. Nagurney, and T. Wakolbinger, "On a paradox of traffic planning," *Transportation Science*, vol. 39, no. 4, pp. 446–450, 2005.

[60] F. Wu, P. Varaiya, P. Spiller, and S. Oren, "Folk theorems on transmission access: Proofs and counterexamples," *Journal of Regulatory Economics*, vol. 10, pp. 5–23, 1996.

5 Infrastructure planning

This chapter zooms out to consider the design of power systems. For example, if a city's power consumption has grown in support of new industry or increased population, should it become the home of a new generator or the destination of a new transmission line? What if there are 100 such choices, all within the same interconnected power system? Additionally, each generation project will take five years to complete and each transmission project ten, over which the energy consumption landscape may change significantly. Finally, each element of the design must respect environmental restrictions, harmonize with a potentially poorly designed legacy system, and be separately financed. It's no wonder that the evolution of power systems looks as much like the product of organic synthesis as deliberate design. It is hence somewhat more reasonable to *plan* power systems and then hope that plans stay consistent with reality to yield efficient and reliable infrastructures.

Even in the absence of the factors listed above, power system planning remains a source of intractable problems because it almost always entails the use of integer variables. Unfortunately, any plan that builds a one-micron-thick transmission line or a two-watt nuclear power plant is not very useful. This makes many power system planning problems NP-hard, and thus very hard to solve. Fortunately, although these problems cannot be solved in polynomial time, powerful mixed-integer programming heuristics like cutting planes and branch-and-bound make moderate-sized problems tractable.

This chapter ignores many of the above real-world details of power system planning and focuses on its underlying mathematical skeleton. This approach is to respect discrete constraints and seek formulations that have convex continuous relaxations, as discussed in Section 2.2.4. More precisely, if the problem we'd like to solve is

$$\begin{aligned} \underset{x}{\text{minimize}} \quad & f(x, y) \\ \text{subject to} \quad & g_i(x, y) \leq 0 \\ & y_i \in \mathbb{Z}, \end{aligned}$$

then we want $f(x, y)$ and each $g_i(x, y)$ to be convex. Of course, this has been our approach to mixed-integer problems all along, but we restate it here because it applies throughout this section. As we will see, most (but not all) of the continuous nonconvexities are power flow constraints, which we relax or approximate using tools from Chapters 2 and 3. Consequently, all of our final formulations will be MILPs, MISOCPs, or MISDPs. Once again, we are merely layering atop optimal power flow models.

5.1   Nodal placement and sizing

There are many types of resources that are allocated node-wise over power systems, such as real and reactive power sources and energy storage. This section sketches generic models for this class of problems. The objective is not to comprehensively describe all or most such scenarios, which would be impossible given the range of components and objectives, but rather to illustrate how to conceptually approach them.

The installation of a resource is represented by an integer variable ξ_i, which determines the allowable range of its output, ρ_i. This relationship is mathematically expressed by the constraints

$$\underline{\rho}_i \xi_i \leq \rho_i \leq \overline{\rho}_i \xi_i \tag{5.1}$$

$$0 \leq \xi_i \leq \overline{\xi}_i \tag{5.2}$$

$$\xi_i \in \mathbb{Z}. \tag{5.3}$$

If nothing is installed, $\xi_i = \rho_i = 0$. If $\overline{\xi}_i = 1$, ξ_i is a binary variable representing pure resource placement, and the above are disjunctive constraints as in Example 2.6. If $\overline{\xi}_i > 1$, ξ_i represents both whether the resource exists at all and in what quantity.

Resource placement is essentially always a discrete decision. Sizing, however, can be discrete as described, or continuous if any size is viable or if the discrete options are spaced finely enough that a continuous approximation is justified. In case of the former, ρ_i represents the output of the resource. In case of the latter, ξ_i is purely a binary placement decision, and, if $\xi_i = 1$, ρ_i is both the resource's continuous size and constant output. While an additional variable can be defined to distinguish the output from continuous sizing, it is omitted here because it is conceptually similar.

We assume that cost vectors c and d are respectively associated with ξ and ρ, so that the total installation cost is

$$\sum_i c_i \xi_i + d_i \rho_i.$$

If the goal of the additions is to improve system performance, there is usually an operational cost as well, which we denote $\mathcal{F}$. For example, $\mathcal{F}$ could be fuel costs, $\sum_i f_i(p_i)$, future maintenance costs, or resistive losses, which are minimized via

$$\mathcal{F} = \sum_{ij} p_{ij} + p_{ji}$$

or

$$\mathcal{F} = \sum_i p_i.$$

The convex relaxations of Section 3.3 are especially useful for resistive losses, which are neglected by the linearized models of Section 3.2.

5.1.1   Problem types and greedy algorithms

This section discusses the two generic formats in which resource placement problems are usually specified and then simple greedy heuristics for quickly obtaining suboptimal solutions.

Adding resources to attain feasibility

In this case, the purpose of the additions is to make the system feasible, for example, to accommodate anticipated load increases. The goal is then to find the least expensive set of additions that attains feasibility, sometimes plus an operational measure. Feasibility can depend on various physical requirements such as having adequate real power generation or suitable voltages across a network. The objective is a convex function of ξ, ρ, and any remaining variables of the form

$$\mathcal{F} + \sum_i c_i \xi_i + d_i \rho_i. \tag{5.4}$$

Note that if the system is already feasible, i.e., at least one feasible point exists, the solution is usually to add nothing.

When using any of the power flow approximations from Chapter 3, it is possible that an optimal solution will be infeasible under exact power flow. MILPs, MISOCPs, and MISDPs hence often serve as tractable first planning stages, the solutions to which may need to be reinforced to attain feasibility. This is discussed in more detail for transmission planning in Section 5.2.5.

Allocating resources subject to a budget constraint

The goal is now to achieve the maximum performance benefit within an allowable budget, β. The objective is an operational measure, $\mathcal{F}$, and the installation costs appear in the budget constraint

$$\sum_i c_i \xi_i + d_i \rho_i \leq \beta. \tag{5.5}$$

In this scenario, the system is typically already feasible. Because any addition usually improves performance, an optimal solution will exhaust the budget regardless of which approximation is used.

Greedy algorithms

Greedy algorithms are quick ways to obtain suboptimal but sometimes quite good solutions to resource allocation and many other combinatorial optimization problems. Greedy algorithms first allocate the single resource that provides the largest system benefit, then the resource with the second largest, and so on until the budget has been exhausted. Greedy algorithms trade optimality for efficiency by ignoring the couplings between different decisions; we demonstrate this compromise in Example 5.1. For this reason, they are sometimes referred to as *myopic* algorithms.

Observe that greedy algorithms are more naturally suited to budget-constrained problems than feasibility attainment. This is because a performance-based objective is necessary to rank individual resource placement decisions, and a budget is needed to tell the algorithm when to stop. Greedy algorithms can be applied to feasibility-oriented problems, but some measure of "distance to feasibility" must be defined.

The following is a simple greedy algorithm for pure placement problems. Suppose $\xi \in \{0, 1\}^n$, and let Θ and Γ respectively be the set of unplaced and placed resources. Initialize $\Theta = \{1, \ldots, n\}$ and $\Gamma = \emptyset$.

1) For each $i \in \Theta$:
   a) Set $\xi_i = 1$, $\xi_j = 1$ for all $j \in \Gamma$, and $\xi_j = 0$ for all $j \in \Theta \backslash i$.
   b) Set $\mathcal{F}_i$ to the objective, $\mathcal{F}$, evaluated at ξ.
2) Set $j = \underset{i \in \Theta}{\text{argmin}} \ \mathcal{F}_i$ and remove j from Θ.
3) Set $\xi_i = 1$ if $i = j$ or $i \in \Gamma$, and zero otherwise. If ξ is feasible, i.e., does not exceed the total budget or cause other constraint violations, add j to Γ.
4) If $\Theta = \emptyset$, terminate. Otherwise, go to step one.

At termination, Γ is the set of resources to be placed. Sizing decisions can be determined either after placement decisions or within the above greedy algorithms, for instance within step one's loop. Example 5.1 shows that greedy algorithms like this one can return suboptimal solutions even on very simple problems.

Let us now informally evaluate the complexity of the above algorithm. Suppose that evaluating $\mathcal{F}$ and feasibility for a particular value of ξ respectively requires $\tau_1(n)$ and $\tau_2(n)$ operations. In each iteration, $\mathcal{F}$ is evaluated once for each element of Θ, and feasibility is evaluated once. Because there are n iterations, this results in approximately $n^2 \tau_1(n)/2 + n\tau_2(n)$ operations. In each iteration, the minimum element in a list must be found, adding on the order of $n \log(n)$ operations per iteration. Assuming $\tau_1(n)$ and $\tau_2(n)$ are similar linear functions of n, the worst complexity is the cubic dependence on n induced by the objective function evaluations.

This is far better than the number of operations required to find the optimal solution, which is guaranteed to be an exponential function of n in the worst case due to the NP-hardness of integer programming. One could enhance the performance of greedy algorithms by making them simultaneously evaluate pairs or more generally k-tuples of decisions, but the resulting complexity increase would rapidly degrade the prized efficiency of greedy algorithms. However, there are scenarios for which greedy algorithms are provably optimal or very mildly suboptimal, cf. [1, 2]. Often, as discussed in Section 2.3.2, this can depend on network characteristics like radiality, which we have already seen in Section 3.3.1 to be a key determinant of the complexity of power flow.

Example 5.1 *Suboptimality of greedy algorithms.* This example shows that the greedy algorithm can yield suboptimal solutions even on extremely simple problems. Consider the following three-resource problem:

$$\underset{\xi}{\text{maximize}} \ \ 2\xi_1 + 2\xi_2 + 3\xi_3$$

$$\text{subject to } \xi_1 + 2\xi_2 + 3\xi_3 \leq 3$$
$$\xi_i \in \{0, 1\}.$$

The greedy algorithm immediately exhausts the "budget" in this problem by first selecting $\xi_3 = 1$ because it increases the objective more than any other individual decision. The greedy solution is thus $\xi = [0\ 0\ 1]$. The optimal solution, however, is $\xi = [1\ 1\ 0]$, which is the exact complement of the greedy solution.

This is not to say that the greedy solution is a bad solution; it is only one worse than the optimum. But, by neglecting the combined contribution and cost of the first and second resources, the greedy algorithm misses the true optimum.

The problem at hand is actually an instance of the famous knapsack problem [3]. Because the knapsack problem is NP-hard, we know that no polynomial-time algorithm like a greedy one can find optimal solutions in general. As can be seen in this chapter, knapsack problems are often identifiable as subsets of constraints encountered in power system resource allocation and other combinatorial optimization problems.

5.1.2   Power sources

We now discuss real and reactive power source placement problems. They are presented together because both problem classes are straightforwardly described by the generic formulations from earlier in this section.

Generation

Distributed generation is a paradigmatic shift from large, centralized generators to more numerous, small power sources that serve loads over shorter distances [4, 5]. Planning distributed generation has received more attention than the centralized case, cf. [6–8], possibly because it is a more recent topic and entails simultaneously allocating a larger set of resources. The difference is mostly contextual, as the below mathematical formulation applies to both distributed and conventional generation planning.

FEASIBLE SET 5.1 (Mixed-integer generation placement and sizing) A power flow feasible set from Chapter 3, (5.1)–(5.3), and

$$\underline{p}_i^0 \leq p_i^0 \leq \overline{p}_i^0 \tag{5.6}$$

$$p_i = p_i^0 + \rho_i. \tag{5.7}$$

Constraint (5.6) represents the preexisting resource at node i, and constraint (5.7) the real power output augmented by the new addition. Here $\underline{\rho}_i$ would usually be zero and $\overline{\rho}_i$ the maximum output per unit capacity. Note that (5.1), (5.6), and (5.7) can be consolidated into

$$\underline{p}_i^0 + \underline{\rho}_i \xi_i \leq p_i \leq \overline{p}_i^0 + \overline{\rho}_i \xi_i. \tag{5.8}$$

While slightly more concise, this contributes little to computational efficiency because the number of integer variables is unchanged. For concreteness, the application of this formulation is demonstrated in the following example.

Example 5.2 *Generation placement with the linearized power flow.* In this example, generation is added to a subset of candidate nodes $\mathcal{G} \subset \{1, \ldots, n\}$, with the objective of minimizing the cost of new generation needed to attain feasibility. Power flow is modeled by the linearized approximation of Section 3.2.1, and each generator placement decision by a binary variable ξ_i. The full optimization is given by

$$\underset{p,\theta,\xi}{\text{minimize}} \quad \sum_i c_i \xi_i$$

$$\text{subject to} \quad p_{ij} = b_{ij}(\theta_i - \theta_j)$$

$$\sum_j p_{ij} = p_i$$

$$\underline{p}_i \le p_i \le \overline{p}_i, \quad i \in \{1, \ldots, n\}/\mathcal{G}$$

$$\underline{p}_i^0 + \underline{p}_i \xi_i \le p_i \le \overline{p}_i^0 + \overline{p}_i \xi_i, \quad i \in \mathcal{G}$$

$$|p_{ij}| \le \overline{s}_{ij}$$

$$\xi_i \in \{0, 1\}, \quad i \in \mathcal{G}.$$

Reactive power sources

In AC power systems, most lines are usually more inductive than resistive, and consequently large amounts of reactive power are consumed through transmission [9]. Given the enormous cost of transmission and distribution lines, it would be unacceptable for the bulk of their capacities to be occupied by reactive power for reasons of both efficiency and stability. To offset this issue, reactive power sources are placed along transmission and distribution lines, generally in the form of capacitor banks and more recently power electronic FACTS (flexible AC transmission system) devices such as static VAR compensators and static synchronous compensators [10, 11]. While capacitor placement has been optimized since the 1960s [12, 13], it is the form initiated by Baran and Wu [14] that has become one of the most well-known placement problems in the power system literature [15].

Although reactive power placement is a very different physical problem from generation allocation, the mathematical formulation is essentially the same but for reactive rather than real power quantities.

FEASIBLE SET 5.2 (Mixed-integer reactive power placement and sizing) A power flow feasible set from Chapter 3, (5.1)-(5.3), and

$$\underline{q}_i^0 \le q_i^0 \le \overline{q}_i^0$$

$$q_i = q_i^0 + \rho_i.$$

There are two common types of capacitors used for reactive power support: fixed and switched. Fixed capacitors operate at one setting, providing a constant injection. In this case, $\underline{\rho}_i = \overline{\rho}_i$. Switched capacitors can switch output levels, and so $\underline{\rho}_i < \overline{\rho}_i$. Technically, the different output levels are discretely spaced, which can be represented by the additional integrality constraint $\rho_i = \alpha \zeta_i$, $\zeta_i \in \mathbb{Z}$. However, the spacing between output levels is usually fine enough that the computational gains of a continuous approximation outweigh the loss in modeling accuracy [16].

Reactive power is often placed along radial distribution systems to maintain acceptable voltage profiles and reduce losses. For this reason, it is especially appropriate to pair these problems with SOC relaxations such as Feasible Set 3.8 [17], which was shown in Section 3.3.1 to be exact under certain conditions in radial networks. An MISOCP reactive power placement model for general networks is given in Example 5.3.

While functionally similar, reactive power sources from the FACTS family are usually devoted to system stability at the transmission level [18, 19]. This complicates the attendant planning problem because it is hard to integrate dynamic features like stability and feedback into convex planning models. This is very much a source of open problems.

5.1.3   Multiple scenarios

A solution to any of the above planning problems corresponds to a single operating condition. Real power systems pass through wide ranges of operating conditions, for example due to varying energy usage over a day, week, or year, or intermittency from a renewable energy source. Were it not for the computational complexity added by integer constraints, it would be unquestionably correct to incorporate multiple scenarios and the dynamics linking them in planning models. As computational capabilities increase, this becomes more and more viable.

The basic approach is similar to multi-period optimal power flow, described in Section 4.1. Suppose that there is a collection of relevant scenarios, indexed by k, each with its own power flow variables and feasible set. The placement variables, ξ_i, are constant across all scenarios and so are not indexed by k. If ρ_i is a continuous sizing variable, it is constant as well. If it is a resource output, however, in each scenario we have a distinct ρ_i^k and associated cost d_i^k. Similarly, the operational measure should vary across scenarios and is denoted $\mathcal{F}^k$.

When the ρ_i^k are resource outputs, the objective (5.4) becomes

$$\sum_k \mathcal{F}^k + \sum_i c_i \xi_i + \sum_{i,k} d_i^k \rho_i^k,$$

where $\mathcal{F}^k$ corresponds to the operational cost of scenario k. In the budget constrained version, (5.5) becomes

$$\sum_i c_i \xi_i + \sum_{i,k} d_i^k \rho_i^k \leq \beta.$$

Scenario-based approaches are widely used in planning and other power system optimization problems. As a simple example, the multi-scenario version of the real power placement constraint (5.8) becomes

$$p_i^{0,k} + \underline{\rho}_i^k \xi_i \le p_i^k \le \overline{p}_i^{0,k} + \overline{\rho}_i^k \xi_i,$$

where each p_i^k appears separately in each scenario's power flow feasible set. In Example 5.3, each scenario corresponds to a different load configuration on a load duration curve, and the associated cost coefficients are weighted by the duration of each configuration.

Example 5.3 *Reactive power placement for loss minimization.* This example presents the capacitor placement problem from Baran and Wu [14] as an MISOCP. Reactive power sources are added to a subset of candidate nodes, again denoted $\mathcal{G} \subset \{1, \ldots, n\}$. Our objective now is to reduce losses as much as possible while staying under the maximum budget, β. To account for losses and their dependence on reactive power, we must use a relaxed model; we arbitrarily select the SOC branch flow model of Section 3.3.3 from our collection of power flow relaxations.

The placement is optimized over multiple scenarios representing load configurations, which are indexed by k. Each load configuration is weighted in the objective by its expected duration, w_k, given by the load duration curve [20].

The full optimization is given by

$$\underset{p,q,v,\psi,\xi}{\text{minimize}} \quad \sum_k w^k \sum_{ij} p_{ij}^k + p_{ji}^k$$

$$\text{subject to} \quad \sum_j p_{ij}^k = p_i^k$$

$$\sum_j q_{ij}^k = q_i^k$$

$$\underline{p}_i^k \le p_i^k \le \overline{p}_i^k$$

$$\underline{q}_i^k \le q_i^k \le \overline{q}_i^k, \quad i \in \{1, \ldots, n\}/\mathcal{G}$$

$$\underline{q}_i^{0,k} + \underline{\rho}_i^k \xi_i \le q_i^k \le \overline{q}_i^{0,k} + \overline{\rho}_i^k \xi_i, \quad i \in \mathcal{G}$$

$$\left(p_{ij}^k\right)^2 + \left(q_{ij}^k\right)^2 \le \psi_{ij}^k v_i^k$$

$$p_{ij}^k + p_{ji}^k = r_{ij} \psi_{ij}^k$$

$$q_{ij}^k + q_{ji}^k = x_{ij} \psi_{ij}^k$$

$$v_j^k = v_i^k - 2\left(r_{ij} p_{ij}^k + x_{ij} q_{ij}^k\right) + \left(r_{ij}^2 + x_{ij}^2\right) \psi_{ji}^k$$

$$\left(p_{ij}^k\right)^2 + \left(q_{ij}^k\right)^2 \le \overline{s}_{ij}^2$$

$$\underline{v}_i^2 \le v_i^k \le \overline{v}_i^2$$

$$\xi_i \in \{0, 1\}, \quad i \in \mathcal{G}$$

$$\sum_{i \in \mathcal{G}} c_i \xi_i \le \beta.$$

5.1.4　Energy storage

Energy storage is a key technological enabler of renewable energy. Consequently, numerous storage resources will need to be placed throughout power systems as renewable power production increases. Because energy storage is dynamic, the associated mathematical problem is formulated by creating a scenario for each consecutive time period, now indexed t.

Because placement and sizing pertains to physical rather than virtual (demand response) versions of energy storage, the defining energy storage parameters from Feasible Set 4.1 no longer vary in time. The planning version is then simply the combination of Feasible Set 4.1 and the above placement constraints. Recall that this involves creating a copy of a power flow feasible set for each time period and linking them using Feasible Set 5.3 as in multi-period optimal power flow (see Example 4.2).

FEASIBLE SET 5.3 (Mixed-integer energy storage placement and sizing)

$$s_i^t = \alpha_i s_i^{t-1} + \Delta \left(\eta_i^e u_i^{et-1} - u_i^{it-1}/\eta_i^i \right)$$

$$0 \le s_i^t \le \overline{c}_i \rho_i$$

$$p_i^t = u_i^{it} - u_i^{et}$$

$$0 \le u_i^{et} \le -\underline{p}_i \rho_i$$

$$0 \le u_i^{it} \le \overline{p}_i \rho_i$$

$$\underline{\rho}_i \xi_i \le \rho_i \le \overline{\rho}_i \xi_i$$

$$0 \le \xi_i \le \overline{\xi}_i$$

$$\xi_i \in \{0, 1\}.$$

Here ξ_i is a pure placement decision $\left(\overline{\xi}_i = 1 \right)$, and ρ_i is a continuous sizing variable. This is but one of many possible formulations; for example, one could just as realistically make ξ_i the placement and sizing variable by letting $\overline{\xi}_i > 1$ and replacing ρ_i with ξ_i. Alternatively, capacity and power constraints can be sized separately by creating two continuous sizing variables, ρ_i^1 and ρ_i^2, each subject to a distinct instance of (5.1).

5.2　Transmission expansion

The most basic problem in planning transmission systems is choosing which nodes to connect with transmission lines. There are many reasons for doing so, such as to connect geographically constrained new generation, accommodate increasing demand, replace decommissioned capacity, or profit from large nodal price differences.

Transmission planning is fundamentally a *network design problem*, as commonly encountered in operations research and transportation [21]. While optimal power flow is unquestionably the innermost layer of power system optimization, it is predated by optimal transmission system planning, which began as early as 1926 when the first minimum spanning tree algorithm was developed by Otakar Borůvka for the purpose of electrifying Moravia [22, 23]. To be fair, the models discussed here do build on optimal power flow and capture substantially more detail than the original purely graph-theoretic spanning tree formulations.

There is a commonly quoted and fairly accurate adage that transmission lines cost "$1 million per mile," and that is without the often protracted legislative aspects. Despite the obvious justification for optimization, actual transmission plans are often ad hoc solutions rather than the result of a systematic design process. The reason is complexity, in both the computational and non-computational senses. Environmental, legal, and economic constraints coupled with the fact that transmission projects can take more than ten years to complete (at the end of which the scenario for which they were designed may no longer exist) make them virtually impossible to fully model.

Add to the above difficulties that integer variables are required to model quantized line sizes and that there can be *a lot* of feasible solutions. Previously in Section 5.1, we had no more integer variables than the number of nodes. Now there are potentially as many integer variables as node pairs, $n(n-1)/2$. As we will see, the integer variables are not additive, but multiplicative with the other variables, adding yet another nonconvexity.

While transmission planning is most certainly not hopeless, it does push optimization to its tractable limits. It is not surprising that, beginning with Garver [24] and for the next forty years, nearly all transmission system planning formulations began with either the network flow or linearized approximations from Section 3.2 [25, 26]. Because they are notationally lighter but share the same conceptual underpinnings, we will also begin our discussion with the linearized approximation but subsequently focus on the convex approximations of Section 3.3. Further discussion on this approach is given in Taylor and Hover [27, 28].

The resulting models are idealized snapshots of long-term planning processes. They are useful in that they represent the mathematical core that these many scenarios are layered upon and thus enable their efficient solution. However, no transmission plan produced by a model in this section would be complete until it had been subjected to a host of simulation tests and analyses. In this regard, the following models are intended to provide foundations for subsequent, more detailed planning stages by carving out the most computationally onerous aspects.

5.2.1 Basic approach

A standard transmission planning objective is the total line cost,

$$\sum_{ij} c_{ij}\eta_{ij}, \tag{5.9}$$

where c_{ij} is the incremental cost of lines from node i to j and η_{ij} is the amount of that line built. It is easy to imagine alternate objectives, e.g. operational characteristics such as resistive losses or cost of generation. Because it is the constraints rather than the objective that are challenging in transmission planning, we restrict our focus to (5.9).

The goal is to minimize this objective while allowing power to feasibly traverse the network, as in the feasibility attainment problem described in Section 5.1.1.

As with all planning problems, the first such challenge is discreteness: because we can't build a line one micron thick from Boston to San Francisco, η_{ij} is an integer variable. As before, we regard these as unavoidable but surmountable difficulties, and we seek formulations that have convex continuous relaxations. The complication that distinguishes transmission planning from other planning problems is the dependence of line admittances on η_{ij}. Suppose η_{ij}^0 line units already exist between nodes i and j. After building a new line of thickness η_{ij}, the resulting capacity of ij is

$$\bar{s}_{ij}\left(\eta_{ij}^0 + \eta_{ij}\right),$$

and its admittance is

$$\left(g_{ij} - ib_{ij}\right)\left(\eta_{ij}^0 + \eta_{ij}\right).$$

The existing network is denoted by Ω_0 and the network of candidate lines by Ω. The sets Ω_0 and Ω are illustrated for a simple network in Figure 5.1. In some cases, we allow Ω_0 and Ω to be implied by respectively setting $\eta_{ij}^0 = 0$ for $ij \notin \Omega_0$ and $\bar{\eta}_{ij} = 0$ for $ij \notin \Omega$.

In many power flow models, admittances multiply other variable quantities, creating a new bilinearity that makes the problem nonconvex even in the absence of integer constraints. This new nonconvexity is dealt with using the lift-and-project relaxation method of Sherali and Tuncbilek [29], exactly as in Example 2.7. We proceed by relaxing models containing the approximations from Chapter 3, which may be relaxations themselves. The relaxation produces disjunctive constraints as seen in Example 2.6. In some cases, this introduces no further approximation beyond the basic power flow model.

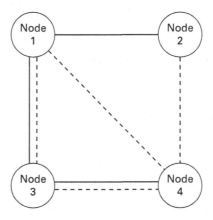

Fig. 5.1   A simple transmission planning scenario. The existing network, Ω_0, is denoted by solid lines, and the network of candidate lines, Ω, by dashed lines. Here, $\Omega_0 = \{12, 13, 34\}$ and $\Omega = \{13, 14, 24, 34\}$.

The story is somewhat different with branch flow formulations, wherein the line impedance is given by

$$\frac{r_{ij} + i x_{ij}}{\eta_{ij}^0 + \eta_{ij}}.$$

In Section 5.2.3, this leads to an approximation that is often simpler and more accurate than voltage-coordinate-based models.

Each of the resulting models can produce infeasible solutions, to which more lines must be added to attain feasibility. In a sense, this is a major shortcoming, but it is also a consequence of the extreme difficulty of transmission planning problems: tractable formulations don't finish the job. Remedies for this issue are discussed in Section 5.2.5.

5.2.2   Linearized models

Traditionally, most transmission planning models have begun with the linearized power flow approximation, as described in Section 3.2.1. We also begin with this model because it is notionally slightly lighter than the relaxed models, but not because it is superior or more practical. The model is obtained by incorporating the line variables η_{ij} into the power flow feasible set, in this case Feasible Set 3.3, as given in Feasible Set 5.4.

FEASIBLE SET 5.4 (Mixed-integer bilinear transmission planning with linearized power flow)

$$p_{ij} = b_{ij} \left(\eta_{ij}^0 + \eta_{ij} \right) (\theta_i - \theta_j) \tag{5.10}$$

$$\sum_j p_{ij} = p_i \tag{5.11}$$

$$|p_{ij}| \le \bar{s}_{ij} \left(\eta_{ij}^0 + \eta_{ij} \right) \tag{5.12}$$

$$\underline{p}_i \le p_i \le \bar{p}_i \tag{5.13}$$

$$0 \le \eta_{ij} \le \bar{\eta}_{ij} \tag{5.14}$$

$$\eta_{ij} \in \mathbb{Z}. \tag{5.15}$$

(5.10) is a nonconvex bilinear constraint, which makes this problem NP-hard even in the absence of integer variables. The lift-and-project relaxation is applied here. We first eliminate the variable p_{ij} through substitution, so that (5.10)-(5.12) are equivalent to

$$\sum_i b_{ij} \left(\eta_{ij}^0 + \eta_{ij} \right) (\theta_i - \theta_j) = p_i \tag{5.16}$$

$$b_{ij}|\theta_i - \theta_j| \le \bar{s}_{ij}, \quad ij \in \Omega_0 \tag{5.17}$$

$$b_{ij}\eta_{ij}|\theta_i - \theta_j| \le \bar{s}_{ij}\eta_{ij}. \tag{5.18}$$

Note that if (5.17) alone was enforced over all the lines, false constraints would arise on angles between nodes that are not directly connected.

Before relaxing these constraints, we first construct a redundant constraint that tightens the resulting relaxation. Let $ij \in \Omega$, and let $\mathcal{P}_{ij}$ be a path from i to j through Ω_0. Summing (5.17) along $\mathcal{P}_{ij}$ and multiplying by b_{ij} yields

$$b_{ij} \left| \theta_i - \theta_j \right| \le M_{ij}, \quad ij \in \Omega, \tag{5.19}$$

where $M_{ij} = b_{ij} \sum_{kl \in P_{ij}} \bar{s}_{kl}/b_{kl}$. Clearly, (5.19) is sharpest when $\mathcal{P}_{ij}$ is the shortest path through the graph induced by Ω_0 with edge weights $\bar{s}_{ij}/b_{ij}$. If no path between i and j exists in Ω_0, $\mathcal{P}_{ij}$ can be set to the longest path through the existing and candidate networks [30] or, even more simply, the set of all edges.

We are now ready to apply the linear lift-and-project relaxation. Introduce the substitution

$$\zeta_{ij} \longleftrightarrow b_{ij} \eta_{ij}(\theta_i - \theta_j),$$

and generate new constraints by substituting ζ into (5.18) and products of (5.17), (5.18), and (5.19); for example, (5.22) is the result of multiplying the upper part of (5.14) with (5.19). The relaxation variables ζ_{ij} are interpretable as the real power flow through the added network described by η_{ij}, $ij \in \Omega$. This yields the relaxation shown in Feasible Set 5.5.

FEASIBLE SET 5.5 (MIL transmission planning relaxation with linearized power flow) (5.13)-(5.15), (5.17), and

$$\sum_j b_{ij} \eta_{ij}^0 (\theta_i - \theta_j) + \zeta_{ij} = p_i \tag{5.20}$$

$$|\zeta_{ij}| \le \min\{M_{ij}, \bar{s}_{ij}\} \eta_{ij} \tag{5.21}$$

$$\left| \zeta_{ij} - b_{ij} \bar{\eta}_{ij} \left(\theta_i - \theta_j \right) \right| \le M_{ij} \left(\bar{\eta}_{ij} - \eta_{ij} \right) \tag{5.22}$$

When $\bar{\eta}_{ij} > 1$, it is possible to tighten the above model by replacing each discrete variable with a collection of binary variables, η_{ij}^k, $k = 1, \ldots, \bar{\eta}_{ij}$. These new variables represent each individual unit of candidate line, and the associated relaxation variables ζ_{ij}^k represent the power flow through each unit. Because the total addition to each line is the sum of the binary variables, the new objective is just a modified version of (5.9),

$$\sum_{ij} c_{ij} \sum_k \eta_{ij}^k.$$

The corresponding feasible set is as follows.

FEASIBLE SET 5.6 (MIL (binary) transmission planning relaxation with linearized power flow) (5.13), (5.17), and

$$\sum_j b_{ij} \eta_{ij}^0 (\theta_i - \theta_j) + \sum_k \zeta_{ij}^k = p_i \tag{5.23}$$

$$\left| \zeta_{ij}^k \right| \le \min\{M_{ij}, \bar{s}_{ij}\} \eta_{ij}^k \tag{5.24}$$

$$\left| \zeta_{ij}^k - b_{ij} \left(\theta_i - \theta_j \right) \right| \leq M_{ij} \left(1 - \eta_{ij}^k \right) \tag{5.25}$$

$$\eta_{ij}^k \in \{0, 1\}, \quad k = 1, \ldots, \overline{n}_{ij}. \tag{5.26}$$

Feasible Sets 5.5 and 5.6 are MILPs that, despite being NP-hard, can often be solved at moderate scales thanks to powerful combinations of branch-and-bound and cutting planes employed by modern solvers. The constraints (5.22) and (5.25) are the disjunctive constraints for which the "disjunctive transmission planning model" is named [31–33]. While the latter binary model is more accurate, it can have substantially more discrete variables than Feasible Set 5.5 and consequently may be prohibitively more expensive computationally.

5.2.3   Branch flow approximation

We begin our coverage of mixed-integer nonlinear models with the branch flow-based approximation from Taylor and Hover [28]. While we could certainly derive relaxed models from the branch flow equations, there is little reason to expect them to perform differently from those in the next section due to the equivalence of the underlying power flow relaxations. Instead, we develop a transmission planning approximation directly from the convex quadratic power flow approximation of Section 3.3.3.

We immediately obtain the following MISOC feasible set by making impedances and line capacities proportional to $\eta_{ij} + \eta_{ij}^0$ in Feasible Set 3.14. Note that here the line sets Ω_0 and Ω are encoded in the parameters η_{ij}^0 and $\overline{n}_{ij}$, conveniently removing the need to distinguish between different sets of lines.

FEASIBLE SET 5.7 (MISOC branch flow transmission planning approximation)

$$\sum_j p_{ij} = p_i \tag{5.27}$$

$$\sum_j q_{ij} = q_i \tag{5.28}$$

$$\underline{p}_i \leq p_i \leq \overline{p}_i \tag{5.29}$$

$$\underline{q}_i \leq q_i \leq \overline{q}_i \tag{5.30}$$

$$p_{ij}^2 + q_{ij}^2 \leq \overline{s}_{ij}^2 \left(\eta_{ij}^0 + \eta_{ij} \right)^2 \tag{5.31}$$

$$r_{ij} \left(p_{ij}^2 + q_{ij}^2 \right) \leq \left(p_{ij} + p_{ji} \right) \left(\eta_{ij}^0 + \eta_{ij} \right) \tag{5.32}$$

$$x_{ij} \left(p_{ij}^2 + q_{ij}^2 \right) \leq \left(q_{ij} + q_{ji} \right) \left(\eta_{ij}^0 + \eta_{ij} \right) \tag{5.33}$$

$$0 \leq \eta_{ij} \leq \overline{n}_{ij} \tag{5.34}$$

$$\eta_{ij} \in \mathbb{Z}. \tag{5.35}$$

Observe that (5.31)-(5.33) are hyperbolic SOC constraints (see Section 2.3.1), and the others are linear. Including discrete line variables has merely changed the CQCP power flow approximation to an MISOCP planning model. Consequently, unlike other transmission planning formulations, this model requires no further modification to attain convexity. This feature should be compared to the preceding linear models, which start with a less accurate approximation and must be subsequently relaxed to attain convexity. As a result, Feasible Set 5.7 is relatively concise compared to other convex transmission planning models. Given the development of advanced MISOCP algorithms and the quality of the underlying quadratically constrained power flow approximation, this formulation is especially practical for high-fidelity transmission planning.

5.2.4   Relaxations

We now discuss MISOC and MISD relaxations of transmission planning models. While notationally more complex, these models admit the same conceptual relaxation as the linearized models in Section 5.2.2, which produce analogous disjunctive constraints. Rather than beginning with full, nonconvex power flow, we augment an SD relaxation, following the approach of Section 5.2.1. We arbitrarily work with the real-valued, voltage-polar coordinate relaxation of Feasible Set 3.11.

FEASIBLE SET 5.8 (MISD bilinear transmission planning) (5.27)-(5.31), (5.34), (5.35), and

$$p_{ij} = \left(\eta_{ij}^0 + \eta_{ij}\right)\left(g_{ij}W_{ii} - g_{ij}W_{ij} + b_{ij}W_{i+n,j}\right) \tag{5.36}$$

$$q_{ij} = \left(\eta_{ij}^0 + \eta_{ij}\right)\left(b_{ij}W_{ii} - g_{ij}W_{i+n,j} - b_{ij}W_{ij}\right) \tag{5.37}$$

$$W_{ij} = W_{i+n,j+n} \tag{5.38}$$

$$W_{i,j+n} = -W_{i+n,j} \tag{5.39}$$

$$\underline{v}_i^2 \le W_{ii} \le \bar{v}_i^2 \tag{5.40}$$

$$W \succeq 0. \tag{5.41}$$

Again, the sets of existing and candidate lines, Ω_0 and Ω, are implicit in η_{ij}^0 and the constraints on η_{ij}. Recall from Section 3.3 that this model would contain an exact description of power flow had we also included the constraint rank $W = 1$. While more complicated than Feasible Set 5.4, all of the above constraints are convex except for the first two, which are bilinear as before. Define

$$W^{ij} = \begin{bmatrix} W_{ii} & W_{ij} & W_{i,i+n} & W_{i,j+n} \\ W_{ji} & W_{jj} & W_{j,i+n} & W_{j,j+n} \\ W_{i+n,i} & W_{i+n,j} & W_{i+n,i+n} & W_{i+n,j+n} \\ W_{j+n,i} & W_{j+n,j} & W_{j+n,i+n} & W_{j+n,j+n} \end{bmatrix},$$

and introduce the relaxation substitution $\Lambda^{ij} \longleftrightarrow \eta_{ij}W^{ij}$. Note that since η_{ij} does not multiply elements of W outside of W^{ij}, there is no need to form bulkier relaxation variables from the products $\eta_{ij}W$. We proceed just as in Section 5.2.2 by taking products of existing constraints and appropriately substituting Λ^{ij}. The resulting relaxation is given in Feasible Set 5.9.

FEASIBLE SET 5.9 (MISD transmission planning relaxation) (5.27)-(5.31), (5.34), (5.35), (5.38)-(5.41), and

$$p_{ij} = \eta_{ij}^0 p_{ij}^0 + p_{ij}^1$$

$$q_{ij} = \eta_{ij}^0 q_{ij}^0 + q_{ij}^1$$

$$p_{ij}^0 = g_{ij} W_{ii} - g_{ij} W_{ij} + b_{ij} W_{i+n,j}$$

$$q_{ij}^0 = b_{ij} W_{ii} - g_{ij} W_{i+n,j} - b_{ij} W_{ij}$$

$$p_{ij}^1 = g_{ij} \Lambda_{11}^{ij} - g_{ij} \Lambda_{12}^{ij} + b_{ij} \Lambda_{32}^{ij}$$

$$q_{ij}^1 = b_{ij} \Lambda_{11}^{ij} - g_{ij} \Lambda_{32}^{ij} - b_{ij} \Lambda_{12}^{ij}$$

$$\Lambda_{kl}^{ij} = \Lambda_{k+2,l+2}^{ij}, \quad k,l \in \{1,2\}$$

$$\Lambda_{k,l+2}^{ij} = -\Lambda_{k+2,l}^{ij}, \quad k,l \in \{1,2\}$$

$$\underline{v}_i^2 \eta_{ij} \le \Lambda_{11}^{ij} \le \overline{v}_i^2 \eta_{ij}$$

$$\underline{v}_j^2 \eta_{ij} \le \Lambda_{22}^{ij} \le \overline{v}_j^2 \eta_{ij}$$

$$\underline{v}_i^2 \left(\overline{\eta}_{ij} - \eta_{ij}\right) \le \overline{\eta}_{ij} W_{ii} - \Lambda_{11}^{ij} \le \overline{v}_i^2 \left(\overline{\eta}_{ij} - \eta_{ij}\right)$$

$$\underline{v}_j^2 \left(\overline{\eta}_{ij} - \eta_{ij}\right) \le \overline{\eta}_{ij} W_{jj} - \Lambda_{22}^{ij} \le \overline{v}_j^2 \left(\overline{\eta}_{ij} - \eta_{ij}\right)$$

$$\left(p_{ij}^0\right)^2 + \left(q_{ij}^0\right)^2 \le \overline{s}_{ij}^2, \quad ij \in \Omega_0$$

$$\left(p_{ij}^1\right)^2 + \left(q_{ij}^1\right)^2 \le \overline{s}_{ij}^2 \eta_{ij}^2, \quad ij \in \Omega$$

$$\left(\overline{\eta}_{ij} p_{ij}^0 - p_{ij}^1\right)^2 + \left(\overline{\eta}_{ij} q_{ij}^0 - q_{ij}^1\right)^2 \le \overline{s}_{ij}^2 \left(\overline{\eta}_{ij} - \eta_{ij}\right)^2, \quad ij \in \Omega_0 \cap \Omega$$

$$\overline{\eta}_{ij} W^{ij} \succeq \Lambda^{ij} \succeq 0.$$

Just as ζ_{ij} represented the real power flow in the added network in Feasible Set 5.5, here the relaxation variables p_{ij}^1 and q_{ij}^1 represent the flow of real and reactive power in the added network. The relaxation variables Λ are similarly relaxed analogs of the matrix variable W. These new variables are related to the non-relaxed variables through multiple disjunctive constraints, such as the second to last one associated with the line capacities.

As with the linearized transmission planning relaxations, efficiency can be traded for accuracy by splitting η_{ij} into $\overline{\eta}_{ij}$ binary variables. However, because of the more complex relationship between power flow and the relaxation variables, we cannot sum along paths to form constraints like (5.19). Whether alternative redundant constraints that tighten these relaxations exist is an open question.

SOC and linear formulations

Although Feasible Set 5.9 is well suited to generic branch-and-bound routines, MISDP does not share the algorithmic maturity of MISOCP or MILP, motivating further, more tractable relaxations. An MISOC transmission planning relaxation is straightforwardly obtained by applying the SOC relaxation of Example 2.9 to all SD constraints, i.e., replacing all SD constraints with nonnegativity constraints on all diagonal terms and two-by-two principal minors. Note that this relaxation could also be constructed by developing a transmission planning model around an SOC optimal power flow relaxation such as Feasible Set 3.10.

There are many ways to translate the SOC transmission planning relaxation into a linear one. The simplest is obtained by simply dropping the SOC constraints and enforcing a linear approximation on the line flow limits (5.31), for example

$$|p_{ij}| + |q_{ij}| \leq \sqrt{2}\bar{s}_{ij}(\bar{\eta}_{ij} + \eta_{ij})$$
$$|p_{ij}| \leq \bar{s}_{ij}(\bar{\eta}_{ij} + \eta_{ij})$$
$$|q_{ij}| \leq \bar{s}_{ij}(\bar{\eta}_{ij} + \eta_{ij}).$$

It has been observed that the resulting solutions often match those produced by linearized models of Section 5.2.2. Alternately, sharper linear relaxations with more variables and constraints are obtainable via the polyhedral relaxation of Section 3.5.4.

5.2.5　Feasibility issues

The basic shortcoming of all the transmission planning models we've given is that they can produce infeasible solutions wherein supply cannot satisfy demand under available capacity. To be useful, the solutions must therefore eventually be reinforced. The NP-hardness of transmission planning ensures that any such practical procedure cannot consistently obtain a true optimal solution. Nevertheless, a reasonable heuristic like the following from Taylor and Hover [28] is likely to obtain good, suboptimal solutions.

1 For the given base network η^0, obtain the optimal solution of a relaxation, η^r.
2 Set $\hat{\eta}^0 = \eta^0 + \eta^r$.
3 Solve the exact transmission planning problem (the transmission planning augmentation of Feasible Set 3.1) using $\hat{\eta}^0$ as a virtual base network. Denote the resulting locally optimal, feasible solution η^f.
4 A feasible, possibly suboptimal solution to the original problem is given by the relaxed solution plus the reinforcement, $\eta^r + \eta^f$.

Step three would be solved using a less specialized optimization method, e.g., nonlinear programming paired with branch-and-bound or a greedy algorithm like in Section 5.1.1, which would likely result in a locally minimal or otherwise suboptimal solution.

Reflecting on this final component of transmission planning computations, it is natural to then question the purpose of relaxations if one must ultimately settle for heuristically obtained solutions. There can be enormous differences in the quality of two feasible, suboptimal solutions, so there is never reason to throw one's hands in the air and abandon all methodological approaches. A relaxed solution may contain a large portion of a very good or optimal solution, effectively pruning the feasible set and shrinking the portion of the problem to be solved with less robust algorithms, thereby reducing the likelihood of obtaining a bad locally minimal solution.

5.3 Summary

Planning problems are fundamentally difficult because they require integer variables. Our approach has been to use the convex approximations of Chapter 3 to ensure that each mixed-integer planning problem has a convex continuous relaxation, enabling the use of advanced heuristics like branch-and-bound and cutting planes. For transmission planning, this required a secondary lift-and-project relaxation following those in Chapter 3.

Presently, only MILPs and some MISOCPs are amenable to these algorithms. However, given the rapidity of MILP's development over the past twenty years and the continued growth of computing power, it is quite reasonable to expect that even MISDPs will admit efficient algorithms in the near future, enabling even more realistic planning formulations.

Here, we have presented rather elemental planning formulations, which can be built up and mixed and matched to accommodate a variety of scenarios. For example, given the locational dependence and intermittency of renewable energy sources, it would be highly appropriate to combine Feasible Sets 5.1 and 5.3 with a transmission planning model from Section 5.2 to obtain a joint generation, storage, and transmission planning model. A model this complex is only meaningful if it can be solved reasonably well, which foremost depends on the tractability of the underlying physical model.

Problems

5.1 Recall from Section 4.1 that energy storage that is grid interfaced through a power electronic converter can provide reactive power, subject to a convex apparent power constraint of the form $p_i^2 + q_i^2 \leq \bar{s}_i^2$. Incorporate this feature into Feasible Set 5.3.

5.2 In the previous problem, now suppose that each energy storage can be matched to any one of a discrete set of converters. Formulate a joint storage-converter planning problem. How is the converter planning problem constrained by the energy storage problem? Suggest a two-stage heuristic in which converters are planned after energy storage has been placed and sized. When might this work well?

5.3 Derive a transmission planning model using the network flow approximation of Feasible Set 3.5. Contrast the model with linearized transmission planning, Feasible Set 5.5.

5.4 Suppose you want to derive a more accurate, higher order SD transmission planning relaxation using lift-and-project, as described in Section 2.3.1. List which constraints from Feasible Set 5.9 to form products of for the next level relaxation (but do not carry out the derivation!).

References

[1] E. Lawler, *Combinatorial Optimization: Networks and Matroids*, ser. Dover Books on Mathematics Series. Dover, 1976.

[2] A. Krause, A. Singh, and C. Guestrin, "Near-optimal sensor placements in Gaussian processes: Theory, efficient algorithms and empirical studies," *Journal of Machine Learning Research*, vol. 9, pp. 235–284, June 2008.

[3] H. Kellerer, U. Pferschy, and D. Pisinger, *Knapsack Problems*. Springer, 2004.

[4] H. L. Willis, *Distributed Power Generation: Planning and Evaluation*. CRC Press, 2000.

[5] T. Ackermann, G. Andersson, and L. Soder, "Distributed generation: a definition," *Electric Power Systems Research*, vol. 57, no. 3, pp. 195–204, 2001.

[6] C. Wang and M. Nehrir, "Analytical approaches for optimal placement of distributed generation sources in power systems," *IEEE Transactions on Power Systems*, vol. 19, no. 4, pp. 2068–2076, 2004.

[7] W. El-Khattam, Y. Hegazy, and M. M. A. Salama, "An integrated distributed generation optimization model for distribution system planning," *IEEE Transactions on Power Systems*, vol. 20, no. 2, pp. 1158–1165, 2005.

[8] A. G. Kagiannas, D. T. Askounis, and J. Psarras, "Power generation planning: A survey from monopoly to competition," *International Journal of Electrical Power & Energy Systems*, vol. 26, no. 6, pp. 413–421, 2004.

[9] A. J. Wood and B. F. Wollenberg, *Power Generation, Operation, and Control*, 3rd ed. Wiley, 2013.

[10] N. G. Hingorani and L. Gyugyi, *Understanding FACTS: Concepts and Technology of Flexible AC Transmission Systems*. Institute of Electrical and Electronics Engineers (New York), 1999.

[11] A. Yazdani and R. Iravani, *Voltage-Sourced Converters in Power Systems*. Wiley–IEEE Press, 2010.

[12] R. Cook, "Optimizing the application of shunt capacitors for reactive-volt-ampere control and loss reduction," *Power Apparatus and Systems, Part III. Transactions of the American Institute of Electrical Engineers*, vol. 80, no. 3, pp. 430–441, 1961.

[13] H. Dura, "Optimum number, location, and size of shunt capacitors in radial distribution feeders: A dynamic programming approach," *IEEE Transactions on Power Apparatus and Systems*, vol. 87, no. 9, pp. 1769–1774, 1968.

[14] M. Baran and F. Wu, "Optimal capacitor placement on radial distribution systems," *IEEE Transactions on Power Delivery*, vol. 4, no. 1, pp. 725–734, Jan. 1989.

[15] H. N. Ng, M. M. A. Salama, and A. Chikhani, "Classification of capacitor allocation techniques," *IEEE Transactions on Power Delivery*, vol. 15, no. 1, pp. 387–392, Jan. 2000.

[16] A. Papalexopoulos, C. Imparato, and F. Wu, "Large-scale optimal power flow: Effects of initialization, decoupling and discretization," *IEEE Transactions on Power Systems*, vol. 4, no. 2, pp. 748–759, May 1989.

[17] S. Abou Jawdeh and R. Jabr, "Mixed integer conic programming approach for optimal capacitor placement in radial distribution networks," in *Universities Power Engineering Conference (UPEC), 2012 47th International*, Sept. 2012, pp. 1–6.

[18] N. Martins and L. T. G. Lima, "Determination of suitable locations for power system stabilizers and static VAR compensators for damping electromechanical oscillations in large scale power systems," *IEEE Transactions on Power Systems*, vol. 5, no. 4, pp. 1455–1469, 1990.

[19] R. Minguez, F. Milano, R. Zarate-Miano, and A. Conejo, "Optimal network placement of SVC devices," *IEEE Transactions on Power Systems*, vol. 22, no. 4, pp. 1851–1860, 2007.

[20] R. Billinton and R. N. Allan, *Reliability Evaluation of Power Systems*, 2nd ed. Springer, 1996.

[21] T. L. Magnanti and R. T. Wong, "Network Design and Transportation Planning: Models and Algorithms," *Transportation Science*, vol. 18, no. 1, pp. 1–55, 1984.

[22] O. Borůvka, "O jistém problému minimálním (About a certain minimal problem)," *Práce Moravské Přírodovědecké Společnosti*, vol. 3, no. 3, pp. 37–58, 1926.

[23] ——, "Příspěvek k řešení otázky ekonomické stavby elektrovodních sítí (Contribution to the solution of a problem of economical construction of electrical networks)," *Elektronický Obzor*, vol. 15, pp. 153–154, 1926.

[24] L. Garver, "Transmission network estimation using linear programming," *IEEE Transactions on Power Apparatus and Systems*, vol. 89, no. 7, pp. 1688–1697, Sept. 1970.

[25] R. Romero, A. Monticelli, A. Garcia, and S. Haffner, "Test systems and mathematical models for transmission network expansion planning," *IEE Proceedings – Generation, Transmission and Distribution*, vol. 149, no. 1, pp. 27–36, 2002.

[26] G. Latorre, R. Cruz, J. Areiza, and A. Villegas, "Classification of publications and models on transmission expansion planning," *IEEE Transactions on Power Systems*, vol. 18, no. 2, pp. 938–946, May 2003.

[27] J. A. Taylor and F. S. Hover, "Linear relaxations for transmission system planning," *IEEE Transactions on Power Systems*, vol. 26, no. 4, pp. 2533–2538, Nov. 2011.

[28] ——, "Conic AC transmission system planning," *IEEE Transactions on Power Systems*, vol. 28, no. 2, pp. 952–959, 2013.

[29] H. D. Sherali and C. H. Tuncbilek, "A global optimization algorithm for polynomial programming problems using a reformulation-linearization technique," *Journal of Global Optimization*, vol. 2, pp. 101–112, 1992.

[30] S. Binato, M. Pereira, and S. Granville, "A new Benders decomposition approach to solve power transmission network design problems," *IEEE Transactions on Power Systems*, vol. 16, no. 2, pp. 235–240, May 2001.

[31] R. Villasana, "Transmission network planning: A method for synthesis of minimum cost secure networks," Ph.D. dissertation, Rensselaer Polytechnic Institute, Troy, NY, 1984.

[32] A. Sharifnia and H. Z. Aashtiani, "Transmission network planning: A method for synthesis of minimum-cost secure networks," *IEEE Transactions on Power Apparatus and Systems*, vol. 104, no. 8, pp. 2025–2034, Aug. 1985.

[33] S. Granville and M. V. F. Pereira, "Analysis of the linearized power flow model in Bender's decomposition," Stanford University, CA, Tech. Rep., 1985, EPRI-report RP 2473-6.

6 Economics

The power system of yore was a jungle. In cities like New York and Chicago, dozens of companies haphazardly strung up their own wires, each carrying its own special blend of electricity. In the United States, Samuel Insull, who began working for Edison in the 1880s, recognized the need for standardization. By the 1920s, he owned a significant share of Chicago's electric power industry, which he made extremely profitable through uniformity, scale, and shady financial practices. His company imploded during the Great Depression, and by the 1930s he had fled to Europe with a villainous reputation. He was acquitted of all charges shortly thereafter but died in 1938 with a debt of nearly $14 million.

Progressive or scoundrel, Samuel Insull's actions paved the way for the Public Utility Holding Company Act of 1935, which effectively made power systems government-regulated, *vertically integrated* monopolies. Over the next forty years, the power system developed into the dependable infrastructure we know today. Concurrently, it became an inefficient, rigid, environmentally hostile dinosaur. In 1978, the U.S. Congress passed the Public Utility Regulatory Policies Act, which forced utilities to purchase power from less expensive *independent power producers* at their avoided costs of generation, marking the first incursion of competition into the electricity sector in decades [1, 2]. In spurts over the next twenty years, the North American electric power system was *deregulated* to the applause of economists and chagrin of many engineers.

Then strange things began to happen, the most notable of which was the California Electricity Crisis [3, 4]. At the center of the crisis was the Enron scandal, and at the center of that, Kenneth Lay, perceived by some as a modern, mirrored version of Samuel Insull. Enron and its ilk gamed every aspect of the vulnerable new electricity markets, leading to extreme price volatility and blackouts. This initial naiveté has been blamed on both economists' failure to understand the physical nature of electric power and engineers' failure to appreciate the complexity and impact of economic systems; in retrospect, one has to wonder if some lessons can only be learned from experience. An excellent historical account of the beginnings of power systems up through deregulation and its pitfalls is given in endnote [5].

Having survived its early missteps, deregulation appears to have gained a relative degree of permanence and at the very least has produced an abundance of interesting technical and policy research problems. As young industries like renewable energy, storage, and demand-side management take off, new problems at the intersection of engineering and economics arise, and the power system must continue to adapt.

At the heart of deregulation is the electricity market, where businesses literally buy and sell electric power. This chapter develops the underlying mathematical skeleton of electricity markets. The reader is referred to endnotes [6–8] for more comprehensive coverage. Rather than building again on top of optimal power flow as in Chapters 4 and 5, here we turn it upside down with Lagrangian duality. This approach, pioneered by Fred Schweppe et al. [9, 10], is strongly related to the classic welfare theorems of microeconomics. This chapter also addresses nonconvexities and market power, two central reasons why these theorems can never fully apply in practice.

6.1 Background

New students of optimization often assume that duality is just a redundant framework for reformulating optimization problems. This couldn't be further from the truth; in addition to forming the basis of a number of algorithms like primal-dual interior point methods, many solutions to optimization problems are practically unimplementable without the interpretation enabled by duality. A notable example from another infrastructure is Internet congestion control, in which Lagrangian duality is used to efficiently allocate rates to users sharing communication channels [11–13]. In deregulated power systems, Lagrangian duality plays a similarly key role by translating centralized optimal power flow solutions into prices that incentivize selfish, independent market participants to behave in a socially optimal way. This is the main focus of this chapter.

Both applications are in the spirit of a general pair of results from microeconomics called the Fundamental Theorems of Welfare Economics, which comprise a theoretical basis for many old and modern ideas in capitalism. It is therefore also worth noting that they are mathematical idealizations that never fully (or sometimes never at all) apply in reality. Nevertheless, when cautiously wielded, they are an indispensable perspective for converting technical solutions into economic implementations.

This section gives brief summaries of Lagrangian duality and its economic interpretation. In-depth coverage of duality and microeconomic theory are respectively found in endnotes [14–18] and [19–21].

6.1.1 Lagrangian duality

Consider the optimization problem

$$P = \min_{x} f(x)$$
$$\text{subject to } g_i(x) \leq 0$$
$$h_i(x) = 0.$$

In the context of duality, this is referred to as the *primal problem*. Its *Lagrangian* is defined as

$$L(x, \alpha, \beta) = f(x) + \sum_i \alpha_i g_i(x) + \sum_i \beta_i h_i(x).$$

α_i and β_i are the *Lagrange* or *dual multipliers* or *dual variables* associated with the constraints $g_i(x) \leq 0$ and $h_i(x) = 0$, respectively. The *dual function* is the unconstrained minimum of the Lagrangian over the primal variables, x,

$$\mathcal{L}(\alpha, \beta) = \min_x L(x, \alpha, \beta).$$

When the objective and constraints are differentiable and convex, the Lagrangian can be obtained by setting the derivative with respect to x to zero, i.e., solving

$$\nabla L(x, \alpha, \beta) = 0,$$

where ∇ is the gradient operator. This condition is referred to as *stationarity*. The Lagrangian is affine in α and β. Because the Lagrange dual function is the minimum over a family of affine functions, it is concave regardless of the properties of the primal problem.

Now suppose that x is feasible for the primal problem, so that $g_i(x) \leq 0$ and $h_i(x) = 0$. Then if $\alpha \geq 0$,

$$\mathcal{L}(\alpha, \beta) \leq L(x, \alpha, \beta) \leq f(x).$$

Because this holds for all feasible x including the minimum, we have that

$$\mathcal{L}(\alpha, \beta) \leq P.$$

This inequality is known as *weak duality*.

To obtain the largest possible lower bound, we define the *dual problem*, or just "dual" for short:

$$D = \max_{\alpha, \beta} \mathcal{L}(\alpha, \beta)$$

subject to $\alpha \geq 0$.

Because the dual is always a concave maximization, it can provide a (relatively) simple way to compute lower bounds for the primal problem.

Weak duality is equivalently stated as $D \leq P$, motivating the definition of *strong duality*:

$$D = P.$$

Strong duality holds, unsurprisingly, when the primal is convex.

Actually, to be precise, strong duality holds when the primal is convex and satisfies a *constraint qualification* condition, which is usually mild. Perhaps the simplest constraint qualification is Slater's condition. Let $\mathcal{I}$ be the set of inequality constraints that are not affine. If there exists an x that is primal feasible and satisfies

$$g_i(x) < 0 \quad \text{for all } i \in \mathcal{I},$$

then Slater's condition is satisfied. This is often referred to as *strict feasibility*, and such an x as an *interior point*. Because equality constraints cannot be strictly satisfied, Slater's condition does not hold if there is a nonlinear equality constraint (in which case the primal is nonconvex). If the primal is convex and Slater's condition holds, so does strong duality. This is a sufficient condition: strong duality may hold even if Slater's condition is not satisfied.

Note that not all constraints need to be dualized. For example, in the optimization

$$\begin{aligned}
\underset{x \in \mathcal{X}}{\text{minimize}} \quad & f(x) \\
\text{subject to} \quad & g_i(x) \le 0 \\
& h_i(x) = 0,
\end{aligned}$$

we can leave out $x \in \mathcal{X}$, in which case the dual function is

$$\mathcal{L}(\alpha, \beta) = \min_{x \in \mathcal{X}} f(x) + \sum_i \alpha_i g_i(x) + \sum_i \beta_i h_i(x).$$

Because the optimal dual variables do not depend on which constraints are dualized, this is merely a notational convention. This is convenient because most convex optimization software automatically produces both the optimal primal and dual variables (when successfully converged), eliminating the need to actually derive duals when numerical answers are desired. That said, it is illustrative to do so for some special classes, given below.

LP duality

Because LPs have only affine inequality constraints, Slater's condition and hence strong duality always hold. Recall the standard form of an LP:

$$\begin{aligned}
\underset{x}{\text{minimize}} \quad & l^T x \\
\text{subject to} \quad & Ax = b \\
& x \ge 0.
\end{aligned}$$

The Lagrange dual function is given by

$$\begin{aligned}
\mathcal{L}(\alpha, \beta) &= \min_x \ l^T x + \beta^T (b - Ax) - \alpha^T x \\
&= \beta^T b + \min_x \ (l - A^T \beta - \alpha)^T x,
\end{aligned}$$

which is equal to $-\infty$ unless $A^T \beta + \alpha = l$. Because we are maximizing, we apply this condition as a constraint to obtain the standard form LP dual:

$$\begin{aligned}
\underset{\alpha, \beta}{\text{maximize}} \quad & \beta^T b \\
\text{subject to} \quad & A^T \beta + \alpha = l \\
& \alpha \ge 0.
\end{aligned}$$

Note that the dual variable α can be eliminated by consolidating the two constraints into $A^T \beta \le l$.

SOCP and SDP duality

As with the positive orthant, the second-order and positive semidefinite cones are *self-dual*, which means that they are their own *dual cones*. This implies that they also have generic dual expressions. Unlike LP, SOCPs and SDPs have constraints that are not affine, meaning that Slater's condition does not trivially hold. Fortunately, the condition is virtually always satisfied by the SOCPs and SDPs encountered in this book.

Recall that an SOCP in standard form is written

$$\underset{x}{\text{minimize}} \ \ l^T x$$

$$\text{subject to} \ \ \|A_i x + b_i\| \leq c_i^T x + d_i.$$

The dual of an SOCP can be derived using scalar multipliers as presented above or with vector multipliers; the latter approach better suits our subsequent purposes. Introduce the vector and scalar multipliers λ_i and μ_i, which in the dual problem must satisfy the SOC constraint $\|\lambda_i\| \leq \mu_i$. The dual function is given by

$$\mathcal{L}(\alpha, \beta) = \underset{x}{\min} \ l^T x + \sum_i \lambda_i^T (A_i x + b_i) - \mu_i \left(c_i^T x + d_i \right)$$

$$= \sum_i \lambda_i^T b_i - \mu_i d_i + \underset{x}{\min} \ \left(l + \sum_i A_i^T \lambda_i - c_i \mu_i \right)^T x,$$

which is equal to $-\infty$ unless $l + \sum_i A_i^T \lambda_i - c_i \mu_i = 0$. As in the LP case, we include this as a constraint to obtain the standard form SOCP dual:

$$\underset{\lambda, \mu}{\text{maximize}} \ \ \sum_i b_i^T \lambda_i - d_i \mu_i$$

$$\text{subject to} \ \ l + \sum_i A_i^T \lambda_i - c_i \mu_i = 0$$

$$\|\lambda_i\| \leq \mu_i.$$

Slater's condition guarantees strong duality if there exists an x for which

$$\|A_i x + b_i\| < c_i^T x + d_i.$$

The standard form primal SDP is given by

$$\underset{x}{\text{minimize}} \ \ l^T x$$

$$\text{subject to} \ \ F_0 + \sum_i F_i x_i \succeq 0.$$

The dual, below, similarly includes a matrix multiplier, Λ:

$$\underset{\Lambda}{\text{maximize}} \ \ -\text{tr} \ F_0 \Lambda$$

$$\text{subject to} \ \ \text{tr} \ F_i \Lambda - l_i = 0$$

$$\Lambda \succeq 0.$$

Now Slater's condition is satisfied if there exists an x for which

$$F_0 + \sum_i F_i x_i \succ 0.$$

Example 6.1 *The Lagrangian dual and SD relaxation of a QCP.* Because Lagrangian duals are convex regardless of primal convexity, they can be used to approximate nonconvex problems, similar to our use of convex relaxations. This is often referred to as *Lagrangian relaxation* because the optimal dual objective bounds the primal from below. As we discussed in Section 2.2.3 and have seen over the course of this book, many important problems are nonconvex QCPs and hence promising candidates for Lagrangian relaxation. In fact, we have been doing this all along: the Lagrangian dual of a QCP is identical to the dual of its SD relaxation, cf. Example 2.8. Therefore, the dual of the dual of a QCP is its SD relaxation, assuming Slater's condition holds. This example derives the dual of a generic QCP and discusses the implications of this relationship.

Consider the possibly nonconvex QCP

$$\underset{x}{\text{minimize}} \quad x^* Q x + c^* x$$

$$\text{subject to} \quad x^* R_i x + r_i^* x + d_i \leq 0.$$

The SD relaxation of this problem is the SDP

$$\underset{x,X}{\text{minimize}} \quad \text{tr } QX + c^* x$$

$$\text{subject to} \quad \text{tr } R_i X + r_i^* x + d_i \leq 0$$

$$\begin{bmatrix} X & x \\ x^* & 1 \end{bmatrix} \succeq 0.$$

Let us now derive the dual of the QCP. Let α_i be the dual multiplier of the quadratic inequality constraint, and define $\tilde{Q} = Q + \sum_i \alpha_i R_i$ and $\tilde{c} = c + \sum_i \alpha_i r_i$. The Lagrangian dual function is given by

$$\mathcal{L}(\alpha) = \underset{x}{\text{min}} \ x^* \tilde{Q} x + \tilde{c}^* x + \sum_i \alpha_i d_i,$$

which is equal to

$$-\frac{\tilde{c}^* \tilde{Q}^{-1} \tilde{c}}{4} + \sum_i \alpha_i d_i$$

if $\tilde{Q} \succeq 0$ and $\tilde{c}$ is in the range of $\tilde{Q}$, and $-\infty$ otherwise. Note that if $\tilde{Q}$ is low rank, the Moore-Penrose pseudoinverse must be used instead of the inverse (see Section 3.4.2). Using the Schur complement (see Example 2.2.2), we can then write the dual as

$$\underset{\alpha,\kappa}{\text{maximize}} \quad -\kappa/4 + \sum_i \alpha_i d_i$$

$$\text{subject to} \quad \begin{bmatrix} \tilde{Q} & \tilde{c} \\ \tilde{c}^* & \kappa \end{bmatrix} \succeq 0$$

$$\alpha \geq 0.$$

This is an SDP, which is the dual of the above SD relaxation.

Because this equivalence holds for all QCPs and optimal power flow is a QCP, we know that its Lagrangian dual is identical to the dual of its SD relaxation, as observed in endnote [22]. This is a nice fact; when we subsequently develop economic price interpretations of dual variables, we'll be guaranteed that prices from the dual of exact,

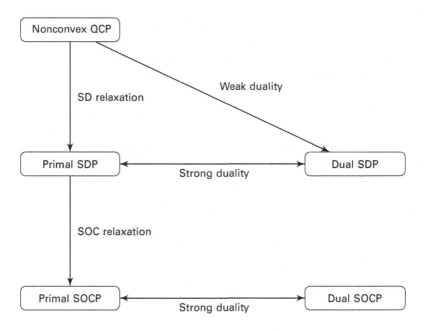

Fig. 6.1 The relationship between nonconvex QCPs and their SDP Lagrangian duals and SD relaxations. For context, the SOC relaxation and its dual are shown as well.

nonconvex optimal power flow are the same as those from its SD relaxation. Of course, this is only useful if the QCP is nonconvex; otherwise, the primal can be far more easily solved as a CQCP, for example, using an SOCP algorithm.

This relationship is illustrated in Figure 6.1, along with the SOC relaxation. Note that the dual of the SOC relaxation is not a relaxation of the dual of the SD relaxation. Rather, the SOC relaxation of the dual SDP has a larger objective value because the dual is a maximization.

Complementary slackness and the Karush-Kuhn-Tucker conditions

Under strong duality, the primal inequality constraints and optimal dual variables satisfy a mutual condition known as *complementary slackness*. Because $D = P$, $f(x) = L(x, \alpha, \beta)$ at the optimal primal and dual solutions, which by definition implies that

$$\sum_i \alpha_i g_i(x) + \sum_i \beta_i h_i(x) = 0.$$

Because x and α are feasible, $h_i(x) = 0$ and $\alpha_i g_i(x) \leq 0$ for each i, which further implies

$$\alpha_i g_i(x) = 0.$$

Intuitively, complementary slackness says that if strong duality holds and an optimal dual variable is positive, the corresponding inequality constraint is active at the optimal primal solution.

For an LP in standard form, complementary slackness reduces to $\alpha_i x_i = 0$. For vector multipliers, this relationship is slightly more complicated. Under strong duality, SOCP and SDP primal and dual variables respectively satisfy

$$\lambda_i^T (A_i x + b_i) - \mu_i \left(c_i^T x + d_i \right) = 0$$

and

$$\text{tr } \Lambda \left(F_0 + \sum_i F_i x_i \right) = 0.$$

Complementary slackness, stationarity, and feasibility together comprise the *Karush-Kuhn-Tucker (KKT) conditions*:

$$\begin{aligned}
\nabla L(x, \alpha, \beta) = 0 && \text{(stationarity)} \\
g_i(x) \leq 0, \ h_i(x) = 0 && \text{(primal feasibility)} \\
\alpha_i \geq 0 && \text{(dual feasibility)} \\
\alpha_i g_i(x) = 0 && \text{(complementary slackness)}
\end{aligned}$$

The KKT conditions are necessary for optimality. If the problem is convex, they are necessary and sufficient, i.e., if satisfied by x, α, and β, then x is primal optimal and α and β are dual optimal.

The optimal dual variables have a more precise meaning than that implied by complementary slackness. Consider the augmented primal problem

$$\begin{aligned}
P = \ & \underset{x}{\text{minimize }} f(x) \\
& \text{subject to } g_i(x) \leq \tau_i \\
& \qquad\qquad h_i(x) = \sigma_i.
\end{aligned}$$

This is equivalent to the original primal problem when $\tau = 0$ and $\sigma = 0$. The optimal dual variables are the sensitivities of the optimal objective P to τ and σ:

$$\alpha_i = - \left.\frac{\partial P}{\partial \tau_i}\right|_{\tau=0, \sigma=0} \quad \text{and} \quad \beta_i = - \left.\frac{\partial P}{\partial \sigma_i}\right|_{\tau=0, \sigma=0}. \tag{6.1}$$

In words, when a constraint is subject to an infinitesimal additive modification, its optimal dual multiplier is the rate at which the optimal objective changes. More to the point, if $f(x)$ represents an economic cost, the dual multipliers then represent the price of the constraint. This is the key technical device for relating optimal power flow solutions to electricity prices.

6.1.2 Pricing and the welfare theorems

We now tie the properties of optimal dual variables to economic concepts. For in-depth treatment of these topics, the interested reader is referred to endnotes [19–21]. Consider the following special case of the primal problem given in Section 6.1.1:

$$\underset{x,y}{\text{minimize }} \sum_i f_i(x_i) \tag{6.2}$$

$$\text{subject to } g_i(y) \leq 0 \tag{6.3}$$

$$x_i = h_i(y). \tag{6.4}$$

As usual, this problem is assumed to be convex. The objective of this problem is additively separable, enabling us to think of each item in the sum as the cost or negative benefit of an agent in a market. x_i is then the quantity of a commodity allocated to agent i, and $f_i(x_i)$ their disutility, for example the cost of producing x_i or the negative benefit derived from x_i. The sum $\sum_i f_i(x_i)$ is called *social welfare*.

The Lagrangian of this problem is given by

$$
\begin{aligned}
L(x, y, \alpha, \beta) &= \sum_i f_i(x_i) + \sum_i \alpha_i g_i(y) + \sum_i \beta_i(h_i(y) - x_i) \\
&= \sum_i f_i(x_i) - \beta_i x_i + \sum_i \alpha_i g_i(y) + \sum_i \beta_i h_i(y).
\end{aligned}
$$

The stationarity condition obtained from differentiating the Lagrangian with respect to x_i is

$$\frac{df_i(x_i)}{dx_i} - \beta_i = 0. \tag{6.5}$$

This also happens to be the result of agent i individually solving

$$\underset{x_i}{\text{minimize }} f_i(x_i) - \beta_i x_i. \tag{6.6}$$

Here, β_i is interpretable as the price agent i pays or is paid for x_i, and (6.6) as its cost minus revenue minimization. If β_i is the optimal dual multiplier, the solution to (6.6) is identical to the socially optimal x_i from (6.2)-(6.4). Prices can be set in this way to align individually and socially optimal behaviors.

This is an extremely general mechanism: even if it is not possible for a central authority to set x_i, there is a price, β_i, which incentivizes each individual to behave in a socially optimal way. The other multipliers similarly price the constraints $g_i(y) \leq 0$, although their precise meanings are context dependent. These prices are often referred to as *shadow prices*. Strong duality says that given convexity and an optimal primal solution (x, y), there always exist multipliers (α, β) that support (x, y). We now informally relate this property to economic concepts.

A feasible solution (x, y) is said to be *Pareto efficient* if there is no other feasible solution (x', y'), such that $f_i(x_i') \leq f_i(x_i)$ for all i and $f_i(x_i') < f_i(x_i)$ for at least one i. At a Pareto efficient solution, no one can improve without making someone else worse off. A triple (x, y, β) is a *competitive (or Walrasian) equilibrium* if it satisfies (6.3)-(6.5).

The *First Fundamental Theorem of Welfare Economics* states that any competitive equilibrium is Pareto efficient. The *Second Fundamental Theorem of Welfare Economics* states that, given a Pareto-efficient solution, one can construct a competitive equilibrium with the same agent utilities. These theorems are often not implemented but merely invoked to theoretically justify the alleged efficiency of capitalistic competition, famously known as Adam Smith's "invisible hand of the market." The critical nature of electric power precludes relying entirely on market forces, and so electricity prices are usually set by a centralized authority using the welfare theorems' close relative, duality.

The relationship between Pareto efficiency and competitive equilibria is similar but not equivalent to strong duality. While an optimal solution to (6.2)-(6.4) is Pareto efficient, the converse need not be true. Moreover, our definition of competitive equilibrium is merely a subset of the KKT conditions, all of which are needed to ensure optimality; endnote [23] discusses the pitfalls of using competitive equilibrium-based pricing instead of duality in electricity markets.

We remark that the phrase "competitive equilibrium" is often used to indicate an optimal outcome, i.e., one that satisfies the KKT conditions. Here, we have introduced narrow versions of Pareto efficiency, competitive equilibria, and the welfare theorems primarily to provide broader economic context. To avoid any further ambiguity, we will henceforth employ a duality-based perspective, and we will refer to optimal outcomes as *economic dispatch*, which we define in Section 6.2.

It is essential to be aware of the pitfalls of duality-based pricing and competitive markets, some of which we list below.

- Convexity is the linchpin of strong duality. If any objective term or constraint in (6.2)-(6.4) is nonconvex, the resulting prices may not support an optimal primal solution. We discuss this issue in the context of electricity markets in Section 6.2.4.
- The decomposition of the centralized, social problem into individual optimizations implicitly assumes that each agent is oblivious to the effects of its action on the overall market. More precisely, we have assumed that each agent is a *price-taker* who is unaware that its choices can affect β_i in (6.6). This assumption is dubious here because many electricity markets are *oligopolies* where a small number of large agents can profit unfairly by exerting *market power*. We use the game theoretic tools of the next section to analyze market power in Section 6.3.
- Agents may not be able to optimally solve (6.6) due to limited time, information, and computing power. This is referred to as *bounded rationality* [24].

These non-ideal scenarios are more often the rule than the exception and may be credited for some of deregulation's early failures and current problems. This is not to say that deregulation was a bad idea or that efficient electricity markets are unattainable; some things have improved since the vertically integrated days. Rather, it is essential to proceed with caution and alertness toward the weaknesses of markets.

6.1.3   Game theory

Game theory is a generalization of optimization wherein multiple agents, called *players*, optimize their individual objectives in anticipation of the other players' decisions. By including players' awarenesses of the broader effects of their decisions, game theory allows us to remove some assumptions from the previous section and examine more realistic scenarios. In particular, game theory can describe what happens when the price-taker assumption of the previous section fails to hold, which we term *market power*. In Section 6.3, we will use these tools to assess electricity market vulnerabilities leading to reduced efficiency. As usual, we introduce the minimal set of concepts necessary, referring the reader to endnotes [25, 26] for more thorough expositions.

A *strategic form game* consists of:

- A set of *players*, which in the present context means agents in a market.
- For each player i, a set of pure strategies, S_i.
- For each player i, a (dis)utility function $u_i(s)$, $s \in \times_i S_i$, which it attempts to minimize over s_i.

Trivially, we see that a strategic form game reduces to ordinary optimization when there is only one player. We will often want to refer to the strategies of a single player and those of all remaining players separately. Standard game theoretic notation streamlines this by using $-i$ to represent all players except i; for example, $s_{-i} \in S_{-i}$ represents the strategies of all players but i, which must reside in the set $S_{-i} = \times_{j \neq i} S_j$.

s is a *pure strategy Nash equilibrium* (PNE) if no single player benefits by deviating. Mathematically, this is expressed by the inequalities[1]

$$u_i(s) \leq u_i(t, s_{-i}) \quad \text{for all } t \in S_i. \tag{6.7}$$

If for each player i, u_i is quasiconvex in s_i and continuous and S_i is compact and convex, a PNE is guaranteed to exist [27–29].[2] An even stronger set of conditions is required for uniqueness [30], but we will not make use of them here.

It is quite common that no PNE exists. The notion of *mixed strategy Nash equilibrium* (MNE) generalizes PNE by allowing players to independently randomize their strategies. Let $\mathcal{M}_i$ denote the set of probability distributions over S_i. The utility that player i derives from the mixed strategy $m \in \times_i \mathcal{M}_i$ is defined as the expectation of its utility over the pure strategies, i.e.,

$$u_i(m) = \int_S u_i(s) m(s) ds.$$

We then say that m is an MNE if, for each player,

$$u_i(m) \leq u_i(t, m_{-i}) \quad \text{for all } t \in \mathcal{M}_i. \tag{6.8}$$

If m attaches a probability of one to a single strategy in S, then it reduces to a PNE. Although seemingly exotic at first glance, MNEs do describe a number of realistic behaviors (like sales), and much milder conditions are necessary to guarantee their existence.

Further to the latter point, if S_i is compact and $u_i(s)$ is continuous for each player, an MNE exists [28]. If the game is *finite*, which is to say that each S_i consists of a finite number of discrete points, an MNE *always* exists [31–33]; it was this result and the equilibrium formulation that enabled it that won John Nash the Nobel Prize in Economics. We will, however, only study continuous games here.

We now look at two classical game theoretic models, both of which were formulated well before the formal introduction of game theory. Each instance is a *duopoly*,

---

[1] It is more common in game theory to use maximizing players because u_i often represents some kind of payoff. To stay consistent with our exposition thus far, we use minimizing players. Mathematically, there is no difference.

[2] A function f is quasiconvex if, for each x and y in its domain, $f(\alpha x + (1 - \alpha)y) \leq \max\{f(x), f(y)\}$ for all $\alpha \in [0, 1]$. Because $\max\{f(x), f(y)\} \geq \alpha f(x) + (1 - \alpha)f(y)$, quasiconvexity generalizes convexity.

in which there are only two players. While duopolies are often the only analytically tractable option, they can also represent a sort of worst-case, least competitive scenario because players are able to exert the most influence on market outcomes. From another perspective, the only thing worse than a duopoly is a monopoly.

Example 6.2 *Cournot duopoly.* This example examines an early model of competition proposed by Antoine Augustin Cournot [34], in which two players compete through their output levels, x_1 and x_2. Their marginal production costs are $c_1 x_1$ and $c_2 x_2$, and the market price is a linear function of the total production, $a - b(x_1 + x_2)$ with $a \geq c_i$, $b > 0$; this is sometimes referred to as an *inverse demand function*. Player i maximizes its net profit via

$$\underset{x_i}{\text{minimize}} \; c_i x_i - (a - b(x_i + x_{-i})) x_i.$$

Because this is a convex quadratic, it can be solved by differentiating, yielding the equilibrium conditions

$$c_i - a + b(2x_i + x_{-i}) = 0.$$

Solving for x, we find that the unique PNE is given by

$$x_i = \frac{a - 2c_i + c_{-i}}{3b}.$$

The equilibrium price is then

$$\frac{a + c_1 + c_2}{3},$$

which can be significantly larger than the marginal prices, c_1 and c_2, if a is big enough.

Example 6.3 *Bertrand duopoly.* Joseph Louis François Bertrand flipped the Cournot setup so that players compete for demand through price, q [35]. Suppose that player i receives all of the demand, D, if $q_i < q_{-i}$ and none otherwise (tie rules can be specified but often are not necessary for analysis), and that both players have the same linear marginal cost of production, c. Then player i's utility is

$$u_i(q) = \begin{cases} (q_i - c)D & \text{if } q_i < q_{-i} \\ 0 & \text{if } q_i > q_{-i} \\ (q_i - c)D/2 & \text{if } q_i = q_{-i}. \end{cases}$$

It is easy to show that the unique PNE is $q_i = c$ for both i. The argument is simply to enumerate all other cases and show that a *profitable deviation* always exists. For example, if $q_i > q_{-i} > c$, then player i profits by changing its price to $q_{-i} - \epsilon$ for any $0 < \epsilon < q_{-i}$. If $q_i = q_{-i} > 0$, it is profitable for either player to *undercut* the other, i.e., reduce the price by a small amount so as to capture all of the demand.

Bertrand competition captures the most efficient market outcome, in that both players make exactly zero profit. We'll see in Section 6.3.3 that adding a detail as basic as production capacity limits can dramatically change the outcome of Bertrand competition.

Cournot and Bertrand competition are often taken to represent the least and most efficient market outcomes, in that the latter yields marginal cost pricing, and the former can dramatically exceed it. Sections 6.3.1 and 6.3.3 will construct more realistic models whose outcomes lie between the two.

Let us now comment on some limitations of game theory. The above examples of Cournot and Bertrand competition are extremely simple, and yet, even slight increases in detail like using more general demand functions or increasing the number of players can bring the very existence of equilibria into question. Of course, we should expect that, because it is a generalization, game theory is less tractable than optimization.

While there is still much to be said about the complexity of game theory, basic results show that even simple games can be hard. For instance, finite two-player games are known to be PPAD-complete [36], which while not as bad as being NP-complete does imply considerable difficulty. The computational complexity of games belongs to a growing field called *algorithmic game theory* [37].

Not only does intractability limit our ability to study games, but it also raises questions about their basic tenets. We have assumed each player or agent to be fully *rational*, which here means each player maximizes a known utility function. This is known as the *expected utility hypothesis* and is rooted in the work of Von Neumann and Morgenstern [38]. But how can ordinary people and organizations attain equilibria that are out of reach to mathematicians and supercomputers; in other words, is anyone actually smart enough to play a game? The answer is obviously no. And yet, game theory has proven incredibly useful for modeling interactions between people and organizations.

The key is to recognize that, whereas all prior material in this book approximates repeatable physical phenomena, here we use game theory to describe phenomena that cannot be precisely described (as is usually the case in economics). Moreover, we don't have to account for every detail because the players in the game don't account for them in their decisions either. To effectively use game theory, we must therefore lessen our goal of achieving the best balance of detail and tractability, and rather try to choose just the right set of ingredients that generates insight into the nebulous realm of market power.

6.2   Electricity markets

We now apply the tools of the previous section to develop price interpretations of optimal power flow dual variables, as pioneered by Schweppe et al. [9, 10]. Let us first reflect on a central implication of the last section. When used as prices, dual variables incentivize market participants to autonomously realize socially optimal solutions. If a detail is absent from the primal problem, it will not be reflected in the participant behaviors incentivized by the dual variable prices. Such inconsistencies can have a number of undesirable repercussions in power systems, including:

- Divergence of actual participant actions from physically and socially good actions.
- Lack of future investment due to missing economic signals. Many modern blackouts are partially attributable to inadequate transmission capacity resulting from lacking financial investment [39, 40].
- Vulnerability to gaming and market abuse, as discussed in Section 6.3.

In power systems, there are far too many details to tractably encapsulate in one optimization model, and some phenomena such as gaming may even be fundamentally unavoidable, cf. [37, 41]. Moreover, making markets too complicated may create new problems even if they are based on realistic models. As such, it is virtually impossible to suppress all undesirable market outcomes; the key is therefore to identify the most important ones and build a model that best balances accuracy and tractability, as has been a theme throughout this book. Consequently, our pricing models will be based on optimization models from previous chapters.

We consider a core component of the electricity marketplace known as the *spot market*. In general spot markets, commodities are bought and sold "on the spot," at spot prices. In a typical power system, daily power production schedules are determined through hourly or similar spot markets, and adjustments are made in faster, minute-scale counterparts. Spot markets play central roles in transacting electric power and facilitate many of the benefits of deregulation, but they are not the only channel for transacting electric power. A large portion of electric power is negotiated ahead of time through *bilateral contracts* between individual producers and consumers. These are often long-term agreements to deliver large amounts of power. In theory, negotiation and trading of bilateral contracts eventually converges to ideal spot market outcomes, although this is difficult to verify in practice because contract payments are usually kept private. In addition to *baseload* real power, a number of supporting commodities or *ancillary services* must be procured, such as backup *reserves* in case of unexpected failures and *regulation* to absorb disturbances. The machinery for buying and selling electricity is extremely complex; more comprehensive expositions can be found in endnotes [6–8, 10].

The central financial quantities in spot markets are the prices at which producers and consumers buy and sell power, which are known as *market clearing prices*, *nodal prices*, or *locational marginal prices*. The latter two are interchangeable and indicate the presence of network constraints, which can lead to different prices at different locations. Nodal power payments alone, however, are insufficient for operating an electricity market, and to be mathematically consistent various transmission payments must also be incorporated. Before discussing such scenarios, let us dissect a simple example.

Example 6.4 *The market clearing price.* Consider economic dispatch with the real power balance from Example 3.2:

$$\underset{p}{\text{minimize}} \ \sum_i f_i(p_i)$$

$$\text{subject to} \ \sum_i p_i = 0. \tag{6.9}$$

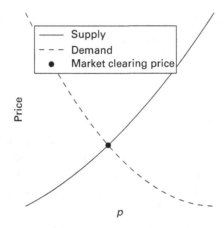

Fig. 6.2   The market clearing price as the intersection of the aggregate supply and demand curves.

Assume that each $f_i(p_i)$ is convex so that strong duality holds. This means that if i is a net producer, $f_i(p_i)$ is its production cost curve, and if i is a net consumer, $f_i(p_i)$ is its demand curve. We formulate the *market clearing price*, which is often characterized as the price where the aggregate supply and demand curves intersect, as shown in Figure 6.2.

The Lagrangian is given by

$$L(p, \lambda) = \sum_i f_i(p_i) - \lambda \sum_i p_i.$$

Differentiating with respect to p, the equilibrium conditions are (6.9) and

$$\frac{\partial f_i(p_i)}{\partial p_i} - \lambda = 0,$$

which matches the sensitivity interpretation of (6.1). When $f_i(p_i) = a_i p_i^2 + b_i p_i + c_i$, this is an equality-constrained quadratic program and can be solved analytically. In this case, the above condition simplifies to $2a_i p_i + b_i = \lambda$, so that the optimal solution is

$$\lambda = \frac{\sum_i b_i/a_i}{\sum_i 1/a_i}$$

$$p_j = \frac{\lambda - b_j}{2a_j}.$$

Suppose now that λ is the optimal multiplier (in the general case). By strong duality, the primal can be solved by minimizing $L(p, \lambda)$, which we can write

$$L(p, \lambda) = \sum_i f_i(p_i) - \lambda p_i.$$

Observe that the sum is separable in i, enabling the minimization to be posed as a collection of individual agent minimizations

$$\underset{p_i}{\text{minimize}} \, f_i(p_i) - \lambda p_i. \qquad (6.10)$$

The first term is the cost (negative benefit) of producing (consuming) p_i, and the second term the payment associated with p_i; thus, λ is the price of p_i. Even if one could not tell a generator how much power to produce or a load how much to consume, there is a price that makes socially optimal levels of production and consumption individually optimal as well.

In this problem, λ is referred to as the *market clearing price* and in fact has been widely used in electricity markets. This makes it all the more unfortunate that it is grossly inaccurate due to the number of unnecessary simplifications in this model. Over the remainder of this section, we will develop more legitimate pricing models, some of which are also currently used by system operators.

In Example 6.4, the agent at node i transacted λp_i for p_i units of real power. Such payments are made to or from a *market operator*, whose job it is to clear markets while ensuring that electricity demand is met in a physically feasible way. Because market operators are usually units within system operators, we will simply use the phrase system operator. A typical spot market for electricity roughly proceeds as follows. Each producer submits a *supply function* that assigns a real power output to each price, the inverse of which times the power output is the production cost curve, $f_i(p_i)$. We remark that "staircase" supply functions have been used to approximate power producers' true costs. This is an unnecessary simplification, which mildly complicates optimal power flow routines and potentially exacerbates market power because staircase functions are discontinuous (and, moreover, not convex but quasiconvex) [42]. Traditionally, *load serving entities* have provided load forecasts to system operators, but many system operators are moving toward more symmetric setups in which loads submit price responsive bids or demand curves [43]. Price responsive loads are sometimes called *elastic*.

On receipt of all agents' supply functions, the system operator solves an optimal power flow in which the objective is the sum of the production costs,

$$\sum_i f_i(p_i).$$

This routine is commonly known as *economic dispatch*.[3] If each generator has been honest, this objective is the total cost of generation. The corresponding dual multipliers are obtained either from the optimal power flow routine or an associated simplified (convex) formulation that has strong duality. Producers are then paid according to the price given by the relevant dual variable, and loads purchase power similarly.[4] Because power in spot markets is not attributable to buyer-seller pairs, spot markets are sometimes called *electricity pools*.

[3] Economic dispatch traditionally refers to minimizing the cost of generation subject to the balance of supply and demand, as in Example 6.4. Since here we primarily consider more realistic models, we let economic dispatch refer to any optimal power flow routine that minimizes operating costs.

[4] At the time of writing, these prices are seen only in wholesale markets. A load in this setting could be a large, single consumer like a manufacturing plant or a load serving entity like an electrical utility that sells power at contracted rates to smaller retail customers like homes and buildings. Exposing retail customers to wholesale prices is discussed in Section 6.2.2.

Before proceeding to the technical formulations, we remark that power systems are not fully or even mostly deregulated; system operators oversee all transactions and scheduling and retain total control over real-time operations. For this reason, they are often referred to as *restructured*. This is necessary for many reasons, the most significant being that electric power is not storable as electricity; real-time imbalances between supply and demand do not simply result in excess inventory or unhappy customers but dynamic instability and physical damage.

6.2.1   Nodal pricing

In Example 6.4, all agents saw the same market clearing price. This price is equivalently interpreted as where the aggregate supply curve crosses the aggregate demand curve. In more realistic models, each agent can see a different price, reflecting its own distinctive constraints such as those describing its location in a transmission network. We now develop such prices by applying the duality-based tools of Section 6.1 in their full generality.

Remember that these prices need not be obtained by solving dual optimizations, which can be tedious to construct. Rather, they are more conveniently obtainable by specifying in a solver's input which primal constraints' dual multipliers should be returned. While optional for the basic models presented here, this approach is quite useful when adding and removing constraints from the more detailed models system operators use in practice.

Pricing linearized approximations

We now construct the pricing scheme obtained from the dual of the linearized power flow of Section 3.2.1, which we reproduce below; see endnotes [10, 23] for a further discussion of pricing with this model. While in some regards this section is just an intermediary between Example 6.4 and the even more realistic SOC and SD models, it highlights some key technical concepts, is widely used by power system operators, and enjoys the high tractability of LP. Linearized economic dispatch is given by:[5]

$$\underset{p,\theta}{\text{minimize}} \sum_i f_i(p_i) \tag{6.11}$$

$$\text{subject to} \tag{6.12}$$

$$\underline{\alpha}_i, \overline{\alpha}_i \geq 0 : \quad \underline{p}_i \leq p_i \leq \overline{p}_i \tag{6.13}$$

$$\lambda_i : \quad p_i = \sum_j b_{ij}(\theta_i - \theta_j) \tag{6.14}$$

$$\chi_{ij} \geq 0 : \quad b_{ij}(\theta_i - \theta_j) \leq \overline{s}_{ij}. \tag{6.15}$$

Dual multipliers are listed to the left of the corresponding constraints in the primal problem. Notice that if $\overline{s}_{ij} = -\underline{p}_i = \overline{p}_i = \infty$, this problem is no different from Example 6.4,

---

[5] In real power systems, there are commonly multiple agents at a single transmission node, each of which sees the same price. To simplify our exposition and because it does not reduce technical generality, we discuss each formulation as though each node houses a single agent.

and a single market clearing price exists. When the constraints are finite and influence the optimal solution, the prices at each node may no longer coincide, hence the phrase nodal prices. A line that is utilized to capacity is called *congested*.

The Lagrangian is given by

$$L(p, \theta, \alpha, \beta, \chi) = \sum_i f_i(p_i) + \underline{\alpha}_i \left(\underline{p}_i - p_i\right) + \overline{\alpha}_i \left(p_i - \overline{p}_i\right)$$

$$+ \lambda_i \left(-p_i + \sum_j b_{ij}(\theta_i - \theta_j)\right) + \sum_j \chi_{ij} \left(b_{ij}(\theta_i - \theta_j) - \overline{s}_{ij}\right)$$

and the stationarity conditions by

$$\frac{\partial f_i(p_i)}{\partial p_i} - \lambda_i - \underline{\alpha}_i + \overline{\alpha}_i = 0 \tag{6.16}$$

$$\sum_j b_{ij}(\lambda_i - \lambda_j + \chi_{ij} - \chi_{ji}) = 0. \tag{6.17}$$

We assume that the rest of the KKT conditions (feasibility and complementary slackness) are also satisfied. The nodal agents face the local optimization problems[6]

$$\underset{p_i}{\text{minimize}} \ f_i(p_i) - \lambda_i p_i \tag{6.19}$$

$$\text{subject to} \ \underline{p}_i \leq p_i \leq \overline{p}_i. \tag{6.20}$$

Here, the nodal price is λ_i. Although it is somewhat tempting here to let market participants truly autonomously realize optimal dispatches by solving (6.19)-(6.20), system operators in real markets dictate producer outputs. In this regard, only load serving entities and active consumers can physically respond to the prices, which otherwise merely provide proper compensation and future investment signals.

χ_{ij} and χ_{ji} are the directional shadow prices associated with line ij. For a given line, at most one of χ_{ij} and χ_{ji} is nonzero due to complementary slackness and because upper

---

[6] The power capacity constraint is not priced. As mentioned in Section 6.1.1, most algorithms will automatically include (6.13) in the Lagrangian, which does not change the optimal dual variables. Indeed, an identical solution is obtained if each agent solves the unconstrained problem

$$\underset{p_i}{\text{minimize}} \ f_i(p_i) - \left(\lambda_i + \underline{\alpha}_i - \overline{\alpha}_i\right) p_i. \tag{6.18}$$

While this does induce the correct decision, we regard it as conceptually incorrect because $\overline{\alpha}_i$ and $\underline{\alpha}_i$ are not power but capacity prices; due to complementary slackness, they are zero if the capacity constraints are passive and represent the value of additional generation capacity when active. Thus, this implementation would spuriously pay agent i to stay within their capacity constraints, which do not represent services and must be respected regardless.

This uncovers an inherent ambiguity in our pricing framework: (6.19)-(6.20) leave the system operator budget balanced once congestion payments are made (by satisfying (6.21)) but can result in non-marginal cost payments to market participants. On the other hand, (6.18) pays each agent at its marginal cost but can leave the system operator with non-zero revenue. We will consider the latter perspective as a tool in Section 6.2.4 for handling discrete nonconvexities.

and lower limits cannot be simultaneously active. Example 6.5 further examines the relationship between nodal and line shadow prices.

Example 6.5 *Nodal price differences and line shadow prices.* The stationarity condition (6.17) has been frequently misinterpreted as implying that the shadow price of a line, $\chi_{ij} - \chi_{ji}$, is equal to the difference of the nodal prices, $\lambda_i - \lambda_j$. This confusion has in part stemmed from network flow models (see Section 3.2.3), which are often used to price transportation systems and for which the assumption is true. The assumption is incorrect for the linearized model at hand, as we demonstrate on the following simple system.

Consider the three-node network depicted in Figure 6.3. Node one generates power at cost $p_1^2/2$, node three consumes power with negative benefit p_3, and node two does not produce or consume power $\left(p_2 = \bar{p}_2 = 0 \right)$, so that the system operator's objective is to minimize $p_1^2/2 + p_3$. Nodes one and three are connected by a line of susceptance b, and the other two lines have susceptance $2b$, so that both paths from node one to three have identical total susceptance, b. If no line constraints bind, the optimal generation and consumption are $p_1 = -p_3 = 1$, all three nodal prices are $\lambda_1 = \lambda_2 = \lambda_3 = 1$, and the line shadow prices are zero.

Now suppose that $\bar{s}_{23} = 1/4$ and $\bar{s}_{12} = \bar{s}_{13} = \infty$. The same amount of power must flow through both paths from nodes one to three because they have the same susceptance. The optimal generation and consumption are therefore $p_1 = -p_3 = 1/2$, with $p_{12} = p_{13} = p_{23} = 1/4$. In this case, by (6.16), $\lambda_1 = 1/2$ and $\lambda_3 = 1$. Because the lines from nodes one to two and one to three are below their capacities, complementary slackness guarantees that their shadow prices are zero. (6.17) requires that $2b(\lambda_1 - \lambda_2) + b(\lambda_1 - \lambda_3) = 0$ and $2b(\lambda_2 - \lambda_1) + 2b(\lambda_2 - \lambda_3 + \chi_{23}) = 0$, which gives $\lambda_2 = 1/4$ and $\chi_{23} = 1$.

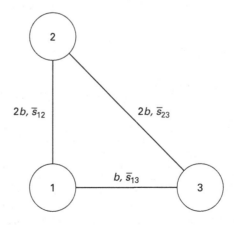

Fig. 6.3 A three-node system. The two paths from nodes one to three have the same susceptances but different capacities, leading to shadow prices that are not equal to nodal price differences.

Hence, even in this very simple network, none of the nodal price differences equals the connecting lines' shadow prices. It is only in radial networks in which the linearized power flow and network flow are identical that nodal price differences always equal the sum of the intermediary shadow prices.

It can be shown from (6.17) and the primal constraints that

$$\sum_i \lambda_i p_i + \sum_{ij} \chi_{ij} \bar{s}_{ij} = 0. \tag{6.21}$$

If any line is congested, the first term is nonzero, leaving the system operator with nonzero profit. To balance its budget, $(\chi_{ij} + \chi_{ji})\bar{s}_{ij}$ would have to be paid to some entity associated with line ij; we discuss mechanisms for implementing this in Section 6.2.3. Observe that because the latter term is always positive, the former is always negative, meaning that the nodal payments always leave the system operator with a positive budget imbalance. Because the budget imbalance arises from congestion, the latter term in (6.21) is often referred to as the congestion rent.

We refer to (6.21) as the *budget equation* for this model because the first term represents the system operator's *budget* or *merchandising surplus*, and the second identifies payments that can balance its budget. One must recall, of course, that the idealizations involved in linearized power flow omit a number of significant cost sources like unit commitment, losses, and contingencies, all of which affect the budget equation and necessitate further consideration. Independent of any particular model, the system operator is said to be *revenue adequate* if its revenues exceed its payments, and *budget balanced* if they are equal.

Many power system operators employ a relative of this section's approach, in which exact power flow (Feasible Set 3.1) is linearized about a desired operating point. This approach has broad applicability because it can integrate a variety of important details like losses, contingency reserves, and reactive power, cf. [44–48] and the preceding references. However, the theory of Section 6.1.2 only heuristically applies to this approach because

- the desired operating point may not be optimal for any model if it is a local minimum, and
- the optimal solution of the resulting LP may not correspond to the linearization point, in spite of the LP's "tangency" to the exact formulation.

Because this approach is mathematically similar to the above, we do not pursue it further here but rather proceed to develop prices for the more accurate relaxations of Section 3.3.

Pricing relaxations

As with LPs, SOCPs and SDPs (usually) have strong duality, enabling us to use the power flow relaxations of Section 3.3 to construct meaningful prices. It must be recalled that since these are relaxations, the optimal solution may not be a physically feasible power flow. However, as discussed in the summary of Chapter 3, this is really a moot criticism because they are more accurate than the linearized approximation, which also produces unphysical power flows and yet has long since gained acceptance as a pricing mechanism. Moreover, as will be seen in the discussion of transmission in the next section, the relaxations seamlessly integrate features like losses that usually must be ad hoc appended to linearized markets. We thus regard prices from relaxations to be at least as valid as those from linearized models and automatically more valid than a single market clearing price from a power balance model (Example 6.4). We first develop this section's concepts with the SOC power flow relaxation and then proceed to the SD relaxation, which is functionally very similar.

Because we later use the voltage-polar coordinate relaxation for the SD case, here we employ the branch flow relaxation (Section 3.3.3) and note that it could be equivalently replaced with any other SOC relaxation from Chapter 3. The branch flow model and its multipliers are shown below.[7]

$$\text{minimize} \atop p,q,v,\psi \quad \sum_i f_i(p_i, q_i) \tag{6.22}$$

$$\text{subject to} \tag{6.23}$$

$$\lambda_i^p : p_i = \sum_j p_{ij} \tag{6.24}$$

$$\lambda_i^q : q_i = \sum_j q_{ij} \tag{6.25}$$

$$\underline{\alpha}_i^p, \overline{\alpha}_i^p \geq 0 : \underline{p}_i \leq p_i \leq \overline{p}_i \tag{6.26}$$

$$\underline{\alpha}_i^q, \overline{\alpha}_i^q \geq 0 : \underline{q}_i \leq q_i \leq \overline{q}_i \tag{6.27}$$

$$\chi_{ij} \geq 0 : p_{ij}^2 + q_{ij}^2 \leq \overline{s}_{ij}^2 \tag{6.28}$$

$$\zeta_{ij} \geq 0 : p_{ij}^2 + q_{ij}^2 \leq \psi_{ij} v_i \tag{6.29}$$

$$\mu_{ij}^p : p_{ij} + p_{ji} = r_{ij}\psi_{ij} \tag{6.30}$$

$$\mu_{ij}^q : q_{ij} + q_{ji} = x_{ij}\psi_{ij} \tag{6.31}$$

$$\delta_{ij} : v_j = v_i - 2(r_{ij}p_{ij} + x_{ij}q_{ij}) + \left(r_{ij}^2 + x_{ij}^2\right)\psi_{ij} \tag{6.32}$$

$$\underline{\gamma}_i, \overline{\gamma}_i \geq 0 : \underline{v}_i^2 \leq v_i \leq \overline{v}_i^2 \tag{6.33}$$

Recall that v_i represents the squared voltage magnitude, $|v_i|^2$, and ψ_{ij} the squared current magnitude, $|I_{ij}|^2$. We can write the convex quadratic and SOC constraints (6.28) and (6.29) in standard SOC form as

$$\left\| \begin{bmatrix} 1 & 0 \\ 0 & 1 \end{bmatrix} \begin{bmatrix} p_{ij} \\ q_{ij} \end{bmatrix} \right\| \leq \overline{s}_{ij}$$

and

$$\left\| \begin{bmatrix} 2 & 0 & 0 & 0 \\ 0 & 2 & 0 & 0 \\ 0 & 0 & 1 & -1 \end{bmatrix} \begin{bmatrix} p_{ij} \\ q_{ij} \\ \psi_{ij} \\ v_i \end{bmatrix} \right\| \leq [0\ 0\ 1\ 1] \begin{bmatrix} p_{ij} \\ q_{ij} \\ \psi_{ij} \\ v_i \end{bmatrix}.$$

---

[7] We have slightly abused the notation by using the symbol "$\geq$" to indicate satisfaction of an SOC constraint.

These constraints appear in the Lagrangian with their respective vector multipliers χ and ν as

$$\chi_{ij}^p p_{ij} + \chi_{ij}^q q_{ij} - \chi_{ij}^s \bar{s}_{ij}$$

and

$$2\zeta_{ij}^p p_{ij} + 2\zeta_{ij}^q q_{ij} + \zeta_{ij}^l \left(\psi_{ij} - v_i\right) - \zeta_{ij}^w \left(\psi_{ij} + v_i\right).$$

If strong duality is attained (e.g., if a Slater point exists), complementary slackness implies that both of these terms are equal to zero. In the dual, χ and ν satisfy the SOC constraints

$$\left\| \begin{bmatrix} \chi_{ij}^p \\ \chi_{ij}^q \end{bmatrix} \right\| \le \chi_{ij}^s \quad \text{and} \quad \left\| \begin{bmatrix} \zeta_{ij}^p \\ \zeta_{ij}^q \\ \zeta_{ij}^l \end{bmatrix} \right\| \le \zeta_{ij}^w.$$

The stationarity conditions are listed below according to the corresponding differentiated variable.

$$p_i: \quad \frac{df(p_i, q_i)}{dp_i} - \lambda_i^p - \underline{\alpha}_i^p + \overline{\alpha}_i^p = 0 \tag{6.34}$$

$$q_i: \quad \frac{df(p_i, q_i)}{dq_i} - \lambda_i^q - \underline{\alpha}_i^q + \overline{\alpha}_i^q = 0 \tag{6.35}$$

$$p_{ij}: \quad \lambda_i^p + \chi_{ij}^p + 2\zeta_{ij}^p - \mu_{ij}^p - \mu_{ji}^p - 2r_{ij}\delta_{ij} = 0 \tag{6.36}$$

$$q_{ij}: \quad \lambda_i^q + \chi_{ij}^q + 2\zeta_{ij}^q - \mu_{ij}^q - \mu_{ji}^q - 2x_{ij}\delta_{ij} = 0 \tag{6.37}$$

$$\psi_{ij}: \quad \zeta_{ij}^l - \zeta_{ij}^w + \mu_{ij}^p r_{ij} + \mu_{ij}^q x_{ij} + \delta_{ij}\left(r_{ij}^2 + x_{ij}^2\right) = 0 \tag{6.38}$$

$$v_i: \quad -\underline{\gamma}_i + \overline{\gamma}_i - \sum_j \zeta_{ij}^l + \zeta_{ij}^w - \delta_{ij} + \delta_{ji} = 0 \tag{6.39}$$

Constraints (6.24) and (6.25) are completely analogous to (6.14), and their multipliers λ_i^p and λ_i^q are respectively the real and reactive power nodal prices. Real and reactive power can thus be priced as with the linearized power flow; in fact, the corresponding real power stationarity condition (6.34) is exactly (6.16).

In the frequent case that (6.27) is strictly satisfied and the objective (6.22) does not depend on reactive power, (6.35) implies that the price of reactive power is zero in the above economic dispatch routine. While mathematically consistent, this outcome inadequately accounts for the cost of reactive power, a problem that is resolved in Example 6.6.

Example 6.6 *Pricing reactive power.* Reactive power plays an integral role in the stable and efficient operation of AC power systems. Intuitively, reactive power should cost substantially less than real power because it does not carry energy and thus does not directly contribute to fuel costs [49]. However, useful prices for reactive power have proven elusive because it is absent from linearized approximations, and the dependence of (6.22) on reactive power is usually very weak, cf. [50–53]. This issue can be partially remedied by incorporating the opportunity costs incurred by providing reactive power, which was discussed in Section 3.5.2. For a generator, this is captured by the (convex quadratic) reactive power capability curve constraints, given in branch flow coordinates by

$$\frac{p_i}{\tan \overline{\theta}_i} \le q_i + \frac{v_i}{x_i} \tag{6.40}$$

$$p_i^2 + q_i^2 \le v_i \overline{I}_i^2 \tag{6.41}$$

$$(v_i + x_i q_i)^2 + x_i^2 p_i^2 \le v_i \overline{E}_i^2. \tag{6.42}$$

If either of the latter two constraints bind, more reactive power means less real power production and hence lower sales. The first constraint can similarly limit real power production if q_i is negative. This coupling between real and reactive power manifests as additional terms in the nodal power balance stationarity conditions, (6.34) and (6.35).

Many newer resources such as energy storage are interfaced to the power system through power electronic converters, enabling them to provide or consume real and reactive power up to a current magnitude limit. In this case, constraint (6.41) alone is appropriate and is written in standard form as

$$\left\| \begin{bmatrix} 2p_i \\ 2q_i \\ v_i - \overline{I}_i^2 \end{bmatrix} \right\| \le v_i + \overline{I}_i^2.$$

We treat this like the other SOC constraints by introducing a new vector multiplier, ϕ, which satisfies the SOC constraint

$$\left\| \begin{bmatrix} \phi_i^p \\ \phi_i^q \\ \phi_i^l \end{bmatrix} \right\| \le \phi_i^w.$$

The new multiplier appears in the Lagrangian as

$$2\phi_i^p p_i + 2\phi_i^q q_i + \phi_i^l \left(v_i - \overline{I}_i^2 \right) - \phi_i^w \left(v_i + \overline{I}_i^2 \right).$$

Differentiating the Lagrangian by p_i and q_i, we obtain augmented relationships for the real and reactive power prices:

$$\frac{df(p_i, q_i)}{dp_i} - \lambda_i^p - \underline{\alpha}_i^p + \overline{\alpha}_i^p + 2\phi_i^p = 0$$

$$\frac{df(p_i, q_i)}{dq_i} - \lambda_i^q - \underline{\alpha}_i^q + \overline{\alpha}_i^q + 2\phi_i^q = 0.$$

The real and reactive nodal prices now depend on each other more strongly through the other constraints on ϕ than when the apparent power constraint is not present.

We conclude this example with a comment on market power. In realistic power systems, real power can be sent over long distances and influences voltage angles system-wide, but reactive power is primarily coupled to the local voltage magnitude. For this reason, reactive power sources usually support local voltage magnitudes. This means that a reactive power source may face little or no competition from providers in other locations and thus has an incentive to raise its price above marginal costs, for example, by misrepresenting its cost function or constraints. Consequently, although reactive power prices are straightforwardly definable, they must also be easy to audit to be practical. For this reason, many reactive power sources are financed through non-price-based techniques.

The primal constraints and stationarity conditions can be used to derive a budget equation as with the linearized model. Specifically, multiplying (6.36) and (6.37) respectively by p_{ij} and q_{ij}, summing them together and over i and j, and making substitutions implied by complementary slackness yield:

$$\sum_i \lambda_i^p p_i + \lambda_i^q q_i + \underbrace{\sum_{ij} \chi_{ij}^s \bar{s}_{ij}}_{\text{Congestion rent}} + \underbrace{\left(2\zeta_{ij}^p - \mu_{ij}^p - \mu_{ji}^p\right) p_{ij}}_{\text{Real power loss charge}}$$

$$+ \underbrace{\left(2\zeta_{ij}^q - \mu_{ij}^q - \mu_{ji}^q\right) q_{ij}}_{\text{Reactive power loss charge}} - \underbrace{2\delta_{ij}\left(r_{ij}p_{ij} + x_{ij}q_{ij}\right)}_{\text{Voltage drop charge}} = 0. \qquad (6.43)$$

As in the linearized case, the first summation represents the revenue accumulated from nodal real and reactive power payments. (6.43) should be compared to the analogous condition for a linearized market, (6.21). The latter only has a congestion term, whereas (6.43) also contains additional terms. If we interpret them in terms of the constraints their multipliers originate from, they represent losses and voltage drop. We henceforth refer to these collectively as loss charges. The budget equation specifies additional payments necessary for the system operator to balance its budget, which we discuss in Section 6.2.3. Note that one could alternatively express (6.43) using the variables v_i and ψ_{ij} instead of p_{ij} and q_{ij} via further substitutions.

We now repeat the above development with the SD relaxation of Section 3.3. Because the SOC models are relaxations of the SD relaxation, prices from the SD relaxation are more physically realistic than any of those above. In fact, as discussed in Example 6.1, the dual of the SD relaxation is the dual of nonconvex, exact optimal power flow, which implies that prices from the SD relaxation are the same as those obtainable from exact formulations.

That said, the SD relaxation's accuracy gain is often very slight, so that prices from an SOC relaxation may be preferable due to the superior tractability of SOCP. Furthermore, recall that the SOC and SD relaxations achieve equivalent accuracy in radial networks due to Theorem 3.1, in which case the SD relaxation should never be used. Here, we arbitrarily use the voltage-polar coordinate version of Section 3.3.2, given below.

$$\underset{p,q,W}{\text{minimize}} \quad \sum_i f_i(p_i, q_i)$$

$$\text{subject to}$$

$$\lambda_i^p : \quad p_i = \sum_j g_{ij} W_{ii} - g_{ij} W_{ij} + b_{ij} W_{i+n,j}$$

$$\lambda_i^q : \quad q_i = \sum_j b_{ij} W_{ii} - g_{ij} W_{i+n,j} - b_{ij} W_{ij}$$

$$\underline{\alpha}_i^p, \overline{\alpha}_i^p \geq 0 : \quad \underline{p}_i \leq p_i \leq \overline{p}_i$$

$$\underline{\alpha}_i^q, \overline{\alpha}_i^q \geq 0: \ \underline{q}_i \leq q_i \leq \overline{q}_i$$

$$\chi_{ij} \succeq 0: \ \left(g_{ij}W_{ii} - g_{ij}W_{ij} + b_{ij}W_{i+n,j}\right)^2$$
$$+ \left(b_{ij}W_{ii} - g_{ij}W_{i+n,j} - b_{ij}W_{ij}\right)^2 \leq \overline{s}_{ij}^2$$

$$\tau_{ij}^1: \ W_{ij} = W_{i+n,j+n}$$

$$\tau_{ij}^2: \ W_{i,j+n} = -W_{i+n,j}$$

$$\underline{\gamma}_i, \overline{\gamma}_i \geq 0: \ \underline{v}_i^2 \leq W_{ii} \leq \overline{v}_i^2$$

$$\Lambda \succeq 0: \ W \succeq 0$$

The vector and matrix multipliers satisfy the SOC and SD constraints

$$\left\| \begin{bmatrix} \chi_{ij}^p \\ \chi_{ij}^q \end{bmatrix} \right\| \leq \chi_{ij}^s \quad \text{and} \quad \Lambda \succeq 0,$$

and, assuming strong duality, the complementary slackness conditions

$$\chi_{ij}^p \left(g_{ij}W_{ii} - g_{ij}W_{ij} + b_{ij}W_{i+n,j}\right) + \chi_{ij}^q \left(b_{ij}W_{ii} - g_{ij}W_{i+n,j} - b_{ij}W_{ij}\right) = \chi_{ij}^s \overline{s}_{ij}$$

and

$$\text{tr} \, \Lambda W = 0.$$

Differentiating the Lagrangian by the primal variables yields the below stationarity conditions:

$$p_i: \ \frac{df(p_i, q_i)}{dp_i} - \lambda_i^p - \underline{\alpha}_i^p + \overline{\alpha}_i^p = 0$$

$$q_i: \ \frac{df(p_i, q_i)}{dq_i} - \lambda_i^q - \underline{\alpha}_i^q + \overline{\alpha}_i^q = 0$$

$$W_{ij}: \ \Lambda_{ji} + \tau_{ij}^1 - g_{ij}\left(\lambda_i^p + \chi_{ij}^p\right) - b_{ij}\left(\lambda_i^q + \chi_{ij}^q\right) = 0$$

$$W_{i+n,j}: \ \Lambda_{j,i+n} + \tau_{ij}^2 + b_{ij}\left(\lambda_i^p + \chi_{ij}^p\right) - g_{ij}\left(\lambda_i^q + \chi_{ij}^q\right) = 0$$

$$W_{i,j+n}: \ \Lambda_{j+n,i} + \tau_{ij}^2 = 0$$

$$W_{i+n,j+n}: \ \Lambda_{j+n,i+n} - \tau_{ij}^1 = 0$$

$$W_{ii}: \ \Lambda_{ii} + \tau_{ii}^1 - \underline{\gamma}_i + \overline{\gamma}_i + \sum_j g_{ij}\left(\lambda_i^p + \chi_{ij}^p\right) + b_{ij}\left(\lambda_i^q + \chi_{ij}^q\right) = 0.$$

Note that a number of the above variables and constraints can be straightforwardly consolidated into simpler expressions by eliminating trivial variables. Manipulating the stationarity conditions and primal constraints yields the budget equation

$$\sum_i \lambda_i^p p_i + \lambda_i^q q_i + \left(\Lambda_{ii} + \Lambda_{i+n,i+n} - \underline{\gamma}_i + \overline{\gamma}_i\right) W_{ii} + \sum_{ij} \chi_{ij}^s \overline{s}_{ij}$$

$$+ \left(\Lambda_{j,i+n} - \Lambda_{j+n,i}\right) W_{i+n,j} + \left(\Lambda_{ji} + \Lambda_{j+n,i+n}\right) W_{ij} = 0. \tag{6.44}$$

Like its linearized and SOC branch flow counterparts (6.21) and (6.43), (6.44) identifies the payments the system operator can make to balance its budget. Here, the revenue surplus from nodal payments and the congestion rents are easily identified, and the remaining terms comprise loss and voltage drop charges similar to those in (6.43).

6.2.2 Multi-period and dynamic pricing

The operational state of an electric power system changes constantly due to time-varying demand and resource availability and component failures or *contingencies*. To maintain feasible operation, resources and device settings must be continually redispatched and updated by rerunning optimal power flow routines. Consequently, the price of electricity varies in time. Traditionally, most electricity consumers have been shielded from this price's volatility by load serving entities like utilities that offer flat or tiered pricing plans. Generators and some larger consumers see this price but often only for the portion of their production or consumption transacted in real-time balancing markets.

Dynamic pricing refers to applying the real-time price of electricity to a significantly greater portion of transactions, cf. [54–56]. Advocates of dynamic pricing claim that it sends the correct market signals and will encourage more sensible energy consumption and proper investment in new infrastructure; for example, high daily and low nightly prices encourage households to perform energy-consumptive tasks like laundry at night, flattening the load curve. In this regard, dynamic pricing is merely a fuller realization of power system deregulation. Opponents point out that it is unrealistic to expect small energy consumers to track a real-time price and that markets are too slow to procure reliability-related services such as frequency and voltage regulation. Worse, improperly implemented dynamic pricing may even cause instability by inducing inaccurate or delayed responses from power consumers [57]. For better or worse, many load serving entities and system operators are moving closer to dynamic pricing but without ceding essential functionalities like frequency and voltage regulation to markets. The most straightforward approach to dynamic pricing is to simply update and apply nodal prices at regular intervals, for example every few minutes.

A distinct but closely related topic is the pricing of dynamic models or, in other words, *multi-period pricing*. In Section 4.1 and again in Section 5.1.4, we discussed applications where successive power flow routines were coupled by dynamic constraints. Such constraints can dramatically affect outcomes and should thus be allowed to influence prices [58–60]. For example, including the constraint (4.4) and Feasible Set 4.1 respectively produce nodal prices that reflect ramping and energy storage constraints. Of course, the ramping constraints themselves should not be priced, as they are the agents' internal constraints, which, like the real power limit (6.13), would leave the system operator budget imbalanced. Note that multi-period pricing need not be implemented as a dynamic price but rather ensures that physical dynamics are reflected by prices at the wholesale level. End consumers could then face either a varying or constant price.

Example 6.7 *Linearized multi-period pricing with ramp constraints.* Consider the following multi-period version of Section 6.2.1's linearized economic dispatch:

$$\underset{p,\theta}{\text{minimize}} \quad \sum_t \sum_i f_i^t \left(p_i^t \right)$$

subject to

$$\underline{\alpha}_i^t, \overline{\alpha}_i^t : \underline{p}_i^t \leq p_i^t \leq \overline{p}_i^t$$

$$\lambda_i^t : p_i^t = \sum_j b_{ij} \left(\theta_i^t - \theta_j^t \right)$$

$$\chi_{ij}^t : b_{ij} \left(\theta_i^t - \theta_j^t \right) \leq \overline{s}_{ij}$$

$$\underline{\iota}_i^t, \overline{\iota}_i^t : \underline{r}_i^t \leq p_i^t - p_i^{t-1} \leq \overline{r}_i^t$$

Recall from Section 4.1.1 that the dynamic ramp constraint limits the amount a resource's real power output can change over one time period.

Like the primal, the dual is composed of a sequence of single period duals strung together by a dynamic constraint. We can see this relationship from the stationarity conditions,

$$\frac{\partial f_i^t \left(p_i^t \right)}{\partial p_i^t} - \lambda_i^t - \underline{\alpha}_i^t + \overline{\alpha}_i^t - \underline{\iota}_i^t + \overline{\iota}_i^t + \underline{\iota}_i^{t+1} - \overline{\iota}_i^{t+1} = 0$$

$$\sum_j b_{ij} \left(\lambda_i^t - \lambda_j^t + \chi_{ij}^t - \chi_{ji}^t \right) = 0.$$

Note that the second condition, which corresponds to differentiating with respect to θ_i^t, is not dynamic because θ_i^t only appears in static constraints. Multiplying it by θ_i^t and summing over i and t, we obtain the multi-period linearized budget condition,

$$\sum_t \sum_i \lambda_i^t p_i^t + \sum_j \chi_{ij}^t \overline{s}_{ij} = 0. \tag{6.45}$$

As in the static case, the budget imbalance identified by the latter term in (6.45) depends only on congestion and not ramping. This implies that even with ramping constraints, nodal and congestion payments alone balance the system operator's budget.

λ_i^t is then the multi-period nodal price for period t. Economic dispatch is attained if the agent at that node solves

$$\underset{p_i}{\text{minimize}} \quad \sum_t f_i^t \left(p_i^t \right) - \lambda_i^t p_i^t \tag{6.46}$$

$$\text{subject to} \quad \underline{p}_i^t \leq p_i^t \leq \overline{p}_i^t \tag{6.47}$$

$$\underline{r}_i^t \leq p_i^t - p_i^{t-1} \leq \overline{r}_i^t \tag{6.48}$$

The local optimization each agent solves at the end of Example 6.7 raises an important issue. In the first period, agent i solves for p_i^t for $t = 1, \ldots, T$ but can only implement p_i^1. Between periods one and two, new information may be acquired that diminishes the quality of the sequence of future prices and decisions, λ_i^t and p_i^t for $t = 2, \ldots, T$. This is because p_i^t and λ_i^t are trajectories that lose validity as the forecasts for which they are optimal deviate from reality.

There are multiple possible approaches to implementing multi-period price trajectories, none of which support economic dispatch in a strict sense but all of which can capture significant multi-period characteristics. One option is to solve for a sequence of prices from the present to a future horizon but then only reveal the current period's prices. Alternatively, an actual price for the current period and price forecasts for future periods could be provided. These are both essentially model predictive control implementations, similar to that discussed in Section 4.1.3. It is difficult to know the best approach, because it is difficult to know the true incentives and decision processes of independent market participants.

Pricing regulation

Thus far, we have developed financial mechanisms for services that fulfill the stationary needs of power systems, such as the steady-state real and reactive power outputs of generators. We now recall our discussion on stability and control in Section 4.2 to address the economics of regulation, that is, the collection of *ancillary services* that maintain smooth operation by absorbing unpredictable forecast errors and disturbances. Historically, the economics of power system regulation have been of deplorably poor design in North America. Up until 2011, frequency regulation providers were paid in proportion to their net expended regulation energy, $\int p\,dt$, which could very well be negative even if a large amount of useful regulation had been provided at significant cost. This practice was eventually deemed inadequate by the Federal Energy Regulatory Commission [61], but the response of regulators was simply to append an additional "mileage" payment of the form $\int |\dot{p}|\,dt$, which, although an improvement, also has no basis in physical or economic modeling.

In this section, we stay the present course and use Lagrange duality to construct a pricing mechanism that supports optimal regulation. One option is to simply recognize that the above multi-period approach admits dynamic modeling if very small or infinitesimal periods are used; this was in fact proposed by Berger and Schweppe [62], not long after Schweppe and colleagues introduced duality-based pricing [9, 10]. In the language of optimal control, this approach's prices are the *costate*, which is the vector of Lagrange multipliers of the dynamic state equation. The result is an optimal price trajectory, which unfortunately loses meaning as a state deviates from its optimal trajectory. Indeed, this is almost guaranteed by the very purpose of regulation: absorbing unpredictable forecast errors and disturbances.

If we further specialize the approach of Berger and Schweppe [62] to the linear quadratic regulator as in endnote [63], we recover a pricing mechanism that accommodates uncertainty. More precisely, the linear quadratic regulator produces an optimal pricing policy in the same sense that it produces an optimal control policy. Furthermore, the pricing policy is optimal under general Gaussian uncertainty due to the certainty equivalence property.

Consider the below special case of the linear quadratic regulation problem (see Section 4.2.2 for the general formulation):

$$\underset{u}{\text{minimize}} \quad \frac{1}{2}\mathbb{E}\left[\sum_i \int_0^T x_i^T Q_{ii} x_i + u_i^T R_{ii} u_i \, dt\right]$$
$$\text{subject to } \dot{x} = Fx + Gu + w, \quad w \sim \mathcal{N}(0, W).$$

Now $x_i^T Q_{ii} x_i$ and $u_i^T R_{ii} u_i$ represent agent i's cost of deviating from nominal operation by x_i and providing regulation u_i, respectively. The objective is separable in i, which allows us to express the general solution, (4.16), agent-wise as

$$u_i = -R_{ii}^{-1} G_i^T P x, \qquad (6.49)$$

where P is again the solution to the matrix Riccati equation, (4.2.2). Now observe that agent i attains (6.49) when faced with the following optimization:

$$\underset{u_i}{\text{minimize}} \; \frac{1}{2} u_i^T R_{ii} u_i - x^T P G_i u_i, \qquad (6.50)$$

where G_i is the ith column block of G. If we regard $u_i^T R_{ii} u_i$ as agent i's cost at a given instant in time, (6.50) looks quite similar to the agent cost optimization (6.19). We thus define $x^T P B_i$ to be the *nodal regulation price* of agent i.

Let us now discuss this payment mechanism. Clearly, the discrete time analog exists, although there are some differences between the two, cf. [63]. The domain of applicability of the nodal regulation price is identical to that of the linear quadratic regulator. For example, zero mean Gaussian uncertainty is a reasonable model for load or renewable fluctuations but not contingencies like the loss of a transmission line. The state x represents deviations to be driven to zero and thus is inappropriate for addressing baseload tasks like load shifting.

6.2.3   Transmission cost allocation

In linearized models, transmission congestion can cause nodal prices to differ, and in relaxations, nodal prices always differ due to congestion and losses. In practice, these price differences can be quite significant. A consequence is that nodal power payments can pose serious risks to power producers and consumers who wish to avoid financial volatility. Additionally, this leaves the system operator budget imbalanced from buying and selling power at different prices, as illustrated in Example 6.8.

More broadly, the economics of transmission should be consistent with power markets, but this does not result from nodal pricing and bilateral contracts alone. One way of viewing why these issues arise is that transmission lines constrain power system operations but are not active market participants, i.e., they do not buy power at one end and sell it at the other.[8] This section discusses two popular tools for linking transmission to power markets, which in turn enable risk-averse market participants to hedge against congestion-induced nodal price volatility.

Transmission rights
Market participants often hedge against nodal price risks by purchasing and trading *transmission rights* ahead of time in forward markets. Transmission rights treat capacity as property, which right owners can rent or use. In some cases, transmission rights

---

[8] For this reason, transmission constraints are sometimes referred to as *negative externalities*. This is a somewhat inverted use of the phrase because transmission is not an exogenous, cost-increasing factor. Rather, transmission reduces costs from the more basic case of no transmission and moreover is under the control of system operators.

are allocated to market participants through auctions, the proceeds of which may then be redistributed to transmission owners or builders. The rights may be subsequently exchanged pairwise between market participants via *bilateral trading*. Market participants can insure themselves against congestion by purchasing transmission rights for lines they use and expect to become congested. A thorough exposition of transmission rights is given in endnote [64].

Defining consistent transmission rights has proven elusive, largely because one cannot choose the path of electric power through a network. For example, a *contract path* is defined as a sequence of lines traversed point-to-point by power sold from a producer to a consumer and specifies the transmission payments associated with a transaction. While common in practice, contract paths are fundamentally flawed: because power does not flow along a single path in a meshed network, payments on the contract path correspond to larger-than-actual power flows, and so-called *loop flows* through off-path lines are never paid for [65]. Physical transmission rights are another type of contract that grants the owner the right to use or allow others to use transmission capacity [23, 66]. Because forward markets cannot perfectly predict real-time power system operations, such rights may specify physically unimplementable outcomes.

In spite of their wide usage in practice, we do not consider contract paths or physical transmission rights because of their above-described physical inconsistencies. The following are two standard types of transmission rights:

Flowgate right (FGR): These entitle the owner to collect payment for power through a single line, assessed at that line's (directional) shadow price.

Financial transmission right (FTR): These entitle the owner to collect payment for power transacted across a node pair, assessed at the nodal price difference.

Unlike physical transmission rights, FGRs and FTRs are purely financial and hence always implementable, regardless of their mathematical or physical consistency.

We now define these rights. Let $\hat{p}_{ij}$ denote the power transfers to which they are applied. The linearized directional rights for a power transfer $\hat{p}_{ij}$ are given by:

Linearized FGR: $\chi_{ij}\hat{p}_{ij}$
Linearized FTR: $(\lambda_i - \lambda_j)\hat{p}_{ij}$

Here χ and λ are the line shadow and nodal prices, as in Section 6.2.1. $\hat{p}_{ij}$, for instance, might be the real power exchange specified by a bilateral contract or a tenth of a line's capacity, $\bar{s}_{ij}/10$. In the case of the latter, the system operator must pay the right owners $\chi_{ij}\hat{p}_{ij} = \chi_{ij}\bar{s}_{ij}/10$, which is nonzero if line ij is congested.

Transmission right owners are often paid by the system operator. If the system operator is required to be budget balanced, the linearized model's budget equation, (6.21), specifies how this can be attained through FGRs. Suppose that line ij has the FGRs $\chi_{ij}\hat{p}_{ij}^k$. Then the system operator's budget balances if

$$\sum_k \hat{p}_{ij}^k = \bar{s}_{ij}$$

for every line in both directions. The system operator is revenue adequate if the above equality is replaced by $\leq$, i.e., they can fulfill their obligation to transmission right owners via revenue surpluses from nodal payments. By parameterizing FGRs by lines' physical capacities, the only future information they depend on is the realized values of the dual multipliers. The application of these concepts is demonstrated in Example 6.8.

Clearly, FGRs and FTRs can be equivalent if defined over the same pair of adjacent nodes, although this is not guaranteed, as would be the case for the system in Example 6.5. Because FTRs may be defined over nonadjacent nodes, they are sometimes referred to as *point-to-point* FTRs. While point-to-point FTRs may not balance the system operator's budget as shown above for FGRs, they are simple to parameterize and a substantial improvement over contract paths.

Example 6.8 *Transmission congestion and financial rights in a linearized two-node network.* Consider a scenario where two nodes that can produce and consume power are connected by a transmission line of capacity $\bar{s}_{12}$. The nodes have demands d_1 and d_2 and can produce power at costs $f_1(p_1) = a_1 p_1^2 + b_1 p_1$ and $f_2(p_2) = a_2 p_2^2 + b_2 p_2$. The system operator solves the cost minimization

$$\underset{p}{\text{minimize}} \ f_1(p_1) + f_2(p_2)$$
$$\text{subject to} \ \ p_1 - p_{12} = d_1$$
$$p_2 + p_{12} = d_2$$
$$|p_{12}| \leq \bar{s}_{12}.$$

In the uncongested case, the optimal power outputs and common nodal price are

$$p_i = \frac{\lambda - b_i}{2a_i}$$

$$\lambda = \frac{\sum_j 2d_j + \frac{b_j}{a_j}}{\sum_j \frac{1}{a_j}}.$$

The power flow from node one through the uncongested line to node two is then given by

$$\tilde{p}_{12} = p_1 - d_1$$
$$= \frac{2(a_2 d_2 - a_1 d_1) + b_2 - b_1}{2(a_1 + a_2)}.$$

Now assume that $a_1 < a_2$ and $b_1 < b_2$, which makes $\tilde{p}_{12}$ positive, and that $\bar{s}_{12} < \tilde{p}_{12}$. The line is now congested so that $p_{ij} = \bar{s}_{ij}$, $p_1 = d_1 + \bar{s}_{12}$, and $p_2 = d_2 - \bar{s}_{12}$. The nodal prices consequently drift apart and become

$$\lambda_1 = 2a_1 (d_1 + \bar{s}_{12}) + b_1$$
$$\lambda_2 = 2a_2 (d_2 - \bar{s}_{12}) + b_2,$$

and the line's shadow price takes on positive value:

$$\chi_{12} = \lambda_2 - \lambda_1 > 0.$$

Note that, as discussed in Section 6.2.1, here the line's shadow price is only equal to the nodal price difference because this is a radial network.

As a result of congestion, producer one is selling $\tilde{p}_{12} - \bar{s}_{12}$ less power than in the uncongested case. Because $d_1 + \bar{s}_{12} < d_2 - \bar{s}_{12}$, $\lambda_1 < \lambda_2$, so that producer one is also selling $\bar{s}_{12}$ at node one at a lower price than it is being purchased for at at node two. This in turn leaves the system operator with the budget surplus $(\lambda_2 - \lambda_1)\bar{s}_{12}$.

Producer one can partially hedge against this profit loss by acquiring the FGR $\chi_{12}\hat{p}_{12}$, where $0 \leq \hat{p}_{12} \leq \bar{s}_{12}$, and the load can similarly avoid buying the power at a higher price than it was sold at by acquiring the remaining FGR $\chi_{12}(\bar{s}_{ij} - \hat{p}_{12})$. When $p_{12} < \bar{s}_{12}$, the line is uncongested and the FGR has no value. When the line becomes congested, the FGR recovers the profit loss due to the price difference up to $\chi_{12}p_{12} = (\lambda_2 - \lambda_1)\bar{s}_{12}$, which is exactly the system operator's budget surplus.

At the time of writing, transmission rights are almost entirely constructed from the linearized model or more detailed linearizations of exact power flow. Whereas transmission rights from the linearized model can only account for the costs of real power congestion, the relaxations can be used to define directional rights for losses and reactive power as well without relinquishing their theoretical properties. Here we define possible FGRs based on the branch flow relaxation, (6.22)-(6.33).

Relaxed real congestion FGR: $\chi_{ij}^p \hat{p}_{ij}$

Relaxed reactive congestion FGR: $\chi_{ij}^q \hat{q}_{ij}$

Relaxed real loss FGR: $\left(2\zeta_{ij}^p - \mu_{ij}^p - \mu_{ji}^p - 2\delta_{ij}r_{ij}\right)\hat{p}_{ij}$

Relaxed reactive loss FGR: $\left(2\zeta_{ij}^q - \mu_{ij}^q - \mu_{ji}^q - 2\delta_{ij}x_{ij}\right)\hat{q}_{ij}$

Observe that the real and reactive loss and congestion FGRs may be decoupled without approximation, enabling various contract bundles to be constructed. We forgo defining analogous rights for the SD relaxation because they may be similarly identified in the SD budget equation, (6.44), and would be functionally very similar to those above. The real and reactive FGRs can be combined as:

Relaxed congestion FGR: $\chi_{ij}^s \hat{s}_{ij}$

Relaxed loss FGR: $\left(2\zeta_{ij}^p - \mu_{ij}^p - \mu_{ji}^p - 2\delta_{ij}r_{ij}\right)\hat{p}_{ij}$
$$+ \left(2\zeta_{ij}^q - \mu_{ij}^q - \mu_{ji}^q - 2\delta_{ij}x_{ij}\right)\hat{q}_{ij}$$

Suppose now that line ij has the congestion FGRs $\chi_{ij}^s \hat{s}_{ij}^k$. As in the above linear case, the system operator pays out the exact congestion charge if

$$\sum_k \hat{s}_{ij}^k = \bar{s}_{ij}.$$

The loss FGRs, however, must be defined as portions of realized power flows, p_{ij} and q_{ij}, to sum exactly to the loss charges, so that

$$\sum_k \hat{p}_{ij}^k = p_{ij} \quad \text{and} \quad \sum_k \hat{q}_{ij}^k = q_{ij}. \tag{6.51}$$

Unlike linearized FGRs, loss FGRs must be parameterized both by realized dual multipliers and power flows to balance the system operator's budget, both of which are unknown at the time of purchase or trade in forward markets.

FTRs for relaxations are harder to define than FGRs because the loss component differentiates sent and received quantities. For example, the real power transfer $\hat{p}_{ij}$ is imprecise in the FTR $\left(\lambda_i^p - \lambda_j^p\right)\hat{p}_{ij}$ because the loss component of the nodal price difference is inconsistently multiplied by a single quantity when the amount sent at i differs from that received at j. As such, we only consider the case of two adjacent nodes:[9]

Adjacent node real FTR: $\lambda_i^p \hat{p}_{ij} + \lambda_j^p \hat{p}_{ji}$
Adjacent node reactive FTR: $\lambda_i^q \hat{q}_{ij} + \lambda_j^q \hat{q}_{ji}$

From (6.36) and (6.37), we can see that a set of these FTRs exactly recovers the budget imbalance in (6.43) if (6.51) is satisfied. We may thus view an FTR defined on adjacent node pairs as the sum of the corresponding congestion and loss FGRs in both directions. Incidentally, this implies that a relation exists between a line's shadow prices and the nodal prices at either end that does not depend on network topology in relaxed models. This is somewhat surprising given that the analogous statement is generally not true in linearized models, as demonstrated in Example 6.5.

Let us now briefly discuss some practical aspects of linearized and relaxed FGRs and FTRs. FTRs are slightly simpler than FGRs in that an agent would have to purchase a collection of FGRs to account for all congested intermediary lines between two nonadjacent nodes. For example, suppose a generator has a long-term bilateral contract with a utility at a nonadjacent node in a non-radial network. In this case, an FTR parametrized by the contracted power quantity and the two nodal prices is considerably simpler than a large portfolio of FGRs. However, FGRs are generally perceived to be more versatile than FTRs because the physical flow through a single line is more clearly defined than that entering and exiting a nonadjacent pair of nodes [66]. Moreover, FTRs do not necessarily lead to budget balance even when assuming a perfect outcome within the linearized model. As such, various approximations are made to prop up FTRs, cf. [67, 68]. On the other hand, FGRs can accommodate the modeling and accuracy advantages of relaxed models like losses and reactive power, which lose consistency in point-to-point FTRs.

---

[9] The two nodal components have the same signs because power flows at opposing line ends have different signs and magnitudes.

Tracing

After power flows have been realized, tracing identifies how much of each nodal contribution is present in each line flow, enabling transmission costs to be decomposed nodally among the market participants [69, 70]. Tracing is not a physical mechanism that can be derived from constitutive modeling but is rather an imposed assumption on causation in transmission networks; justifications for tracing are given in [71, 72]. Tracing can be used as an alternative or complementary technique for allocating transmission costs such as those appearing in the budget equations (6.21) and (6.43) or for more general purposes like recovering the fixed cost of construction or transmission between different system operators.

The main technical concept behind tracing is the so-called *proportional sharing rule*, which relates the composition of incoming and outgoing power flows at each node [69]. Here we derive tracing for real and reactive power flows. Suppose that p_{ij} and q_{ij} comprise a power flow solution, which we want to decompose into nodal contributions. Define the following four node sets:

$\mathcal{I}_i^p$: Nodes sending real power to i, i.e., $p_{ji} > 0$ and $p_{ij} < 0$ for all $j \in \mathcal{I}_i^p$.
$\mathcal{I}_i^q$: Nodes sending reactive power to i.
$\mathcal{O}_i^p$: Nodes receiving real power from i.
$\mathcal{O}_i^q$: Nodes receiving reactive power from i.

Let p_{ij}^k q_{ij}^k be the real and reactive power from generator k in line ij, and let $\mathbb{I}_{ik}$ be an indicator function, which is one if $i = k$ and zero otherwise. The proportional sharing rule states that

$$p_{il}^k = p_{il} \frac{\mathbb{I}_{ik} p_i^+ - \sum_{j \in \mathcal{I}_i^p} p_{ij}^k}{p_i^+ - \sum_{j \in \mathcal{I}_i^p} p_{ij}} \quad \text{for each } l \in \mathcal{O}_i^p \tag{6.52}$$

$$q_{il}^k = q_{il} \frac{\mathbb{I}_{ik} q_i^+ - \sum_{j \in \mathcal{I}_i^q} q_{ij}^k}{q_i^+ - \sum_{j \in \mathcal{I}_i^q} q_{ij}} \quad \text{for each } l \in \mathcal{O}_i^q. \tag{6.53}$$

Note that we are following the convention that, if a node's demand exceeds its supply, the power it produces does not enter the transmission network. The proportional sharing rule is illustrated in Figure 6.4.

It can be extended to include losses by regarding the losses as an extraction at a virtual midpoint node, which yields

$$\frac{p_{ij}^k}{p_{ij}} = \frac{p_{ji}^k}{p_{ji}} \quad \text{and} \quad \frac{q_{ij}^k}{q_{ij}} = \frac{q_{ji}^k}{q_{ji}}. \tag{6.54}$$

Because $p_{ij} = \sum_k p_{ij}^k$ and $q_{ij} = \sum_k q_{ij}^k$, each power flow can be separated into nodal components. (6.52)-(6.54) thus defines an algorithm that can be applied to determine the nodal contributions of the set of power producers. Observe that this tracing rule is applicable to all of the power flow models in Chapter 3, including those feasible under exact power flow, Feasible Set 3.1. The same algorithm can be applied to determine the nodal contributions of a set of consumers by swapping the $\mathcal{I}$ and $\mathcal{O}$ sets and using

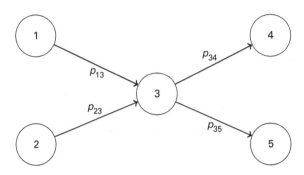

Assume nodes one and two are producers and that the network is lossless. Applying the proportional sharing rule to the flows departing node three yields the decomposition
$$p_{34}^1 = p_{34}\frac{p_{13}}{p_{13}+p_{23}}, p_{34}^2 = p_{34}\frac{p_{23}}{p_{13}+p_{23}}, p_{35}^1 = p_{35}\frac{p_{13}}{p_{13}+p_{23}}, \text{ and } p_{35}^2 = p_{35}\frac{p_{23}}{p_{13}+p_{23}}.$$

nodal power extractions instead of injections, p_i^- and q_i^-. The exact ratio that should be attributed to consumers versus producers is a separate issue decidable only through exogenous notions such as fairness.

6.2.4   Pricing under nonconvexity

The key property linking prices to economic dispatch is strong duality, which rests on convexity. Without convexity, the foundations laid in Section 6.1.1 crumble. We know from Chapter 3 and earlier in this chapter that the power flow constraints are not convex but that convex approximations and markets based on them can be highly functional. Discrete constraints are another type of nonconvexity that more severely compromises the underlying assumptions of competitive markets (economists often refer to this as *lumpiness*). In particular, unit commitment routines such as in Section 4.3 are integral to power system operation and thus should be reflected in power markets. This is problematic, because whereas power flow nonconvexities have convex approximations, integer programs are usually left nonconvex and solved using heuristics like branch-and-cut, making little room for Lagrangian duality-based prices (except in the highly unlikely case that the optimal continuous solution happens to be integer). Furthermore, recalling the unit commitment objective in Section 4.3, there are significant costs associated with discrete startup and shutdown decisions that must be accounted for but are not captured by the continuous economic dispatch routines earlier in this section.

At the time of writing, there is no mathematically consistent way to solve this problem completely, and, moreover, no such solution is to be expected in the form of linear prices. Consequently, power producers often face nodal prices that do not support the optimal dispatch, so that system operators must make additional *make-whole* or *uplift payments* to fully compensate their costs. This section surveys some existing techniques for salvaging prices in the presence of integer constraints.

Most real power markets simply take generator status commitments as fixed and price the resulting convex problem, typically following Example 6.4 or Section 6.2.1 and ad hoc allocating the resulting uplift. A prominent approach in the literature first proposed in endnote [73] fixes integer variables at their optimal values and prices the remaining convex problem. Specifically, consider the optimization

$$\begin{aligned} \underset{x,y}{\text{minimize}} \quad & f(x,y) \\ \text{subject to} \quad & g(x,y) \le 0 \\ & y_i \in \mathbb{Z}. \end{aligned}$$

Suppose that f and g are convex and that y^* is the optimal integer solution. Then consider the convex problem

$$\begin{aligned} \underset{x,y}{\text{minimize}} \quad & \sum f(x,y) \\ \text{subject to} \quad & g(x,y) \le 0 \\ & y = y^*. \end{aligned}$$

Let π be the multiplier of the constraint $y = y^*$. The uplift payment is then distributed through $\pi^T y$, i.e., if y_i is the binary commitment decision of generator i, $\pi_i y_i$ is their payment for this action.

While potentially effective, this approach has some pitfalls, as identified in endnote [74]. The prices π can be both positive and negative and tend to be highly volatile, making seemingly unfair realizations common. A second issue is that they can leave the system operator revenue inadequate, in that the additional payments violate the equalities (6.21) and (6.43); also see the footnote on page 149 for further discussion.

Convex hull pricing is an alternate approach based on the fact that Lagrangian duals are concave regardless of primal convexity [75–77]: simply use the prices produced by dualizing unit commitment. It has been shown that these prices minimize uplift but, of course, do not eliminate it. This approach is additionally intriguing in the present context; recall that Example 6.1 said that the SD relaxation of a QCP is the dual of its Lagrangian dual, and Section 2.2.4 showed that any integer program is expressible as a QCP. Therefore, one can think of convex hull pricing as using the dual multipliers of the SD relaxation of an integer program written as a QCP.

Observe that our discussion on nonconvexities stemming from unit commitment places renewable energy producers and energy storage in a positive light. In particular, these resources have negligible startup and shutdown constraints relative to conventional generators, potentially simplifying unit commitment routines down to multi-period optimal power flow. Markets with a large fraction of renewable and storage resources could therefore have substantially smaller uplifts. However, it remains to be seen if renewables will add new sources of nonconvexity, for example, through chance-constrained or robust formulations of wind and solar variability.

6.3 Market power

In the last section, we developed a powerful toolset for financially incentivizing decentralized, self-interested market participants to realize centrally optimal power system operations. The entire approach rests on a few stringent assumptions introduced in Section 6.1.2, which, as promised, are never actually satisfied in electricity (or most other) markets. One major reason is nonconvexity, which we discussed in Section 6.2.4. We now discuss what is arguably the other major reason, market power. Whereas non-convexity stems from physical elements that cannot be tractably modeled, market power is a behavioral phenomenon and may to some extent be insuppressible.

In a broad sense, an agent has market power when it is able to profit more than normal. In electricity markets, this translates to profiting above what would result from economic dispatch and duality-based nodal pricing with each agent's true cost curve. Market power often stems from agents not behaving like price-takers, which is to say perceiving the effect of their actions on market prices. For example, an agent may be aware of its decision's effect on nodal prices, which means not taking λ as completely exogenous in (6.19). Whereas there is one right way for a market participant to comply with the assumptions of Section 6.1.2, there are an infinity of ways to deviate from them, making market power extremely hard to model in a comprehensive way. Consequently, no one approach to market power has emerged as universally correct, and so our coverage here is somewhat shallower than for other topics in this book.

Game theory is the predominant approach modeling market power. As discussed in Section 6.1.3, game theory often leaves much to be desired in terms of realism and tractability. On the other hand, game theory has been more effective than any other technique at modeling strategic behavior, which cannot be straightforwardly approached from physical principles; one might even venture to say that it is the only game in town. This section presents a few popular models of strategic behavior in electricity markets, starting with the following simple example.

Example 6.9   *Load shifting with energy storage.* Consider a market in which a time-varying, inelastic load $\delta(t)$, $t = 1, \ldots, T$, is served by generation with convex quadratic real power cost, $f(p) = \frac{a}{2}p^2 + bp$. The market clearing price is defined as

$$\lambda(t) = \frac{df(p)}{dp}$$
$$= ap + b.$$

A collection of N energy storages shifts load by buying energy when the market price is low and selling when it is high; this is also known as *arbitrage* and both profits the storages and benefits the power system by reducing generation costs. We model each storage by forcing the energy exchanges, $s_i(t)$, to add to zero over time. The real power demand at a given time is then $\delta(t) - \sum_{i=1}^{N} s_i(t)$, which the generation meets exactly.

Given control over all storages, the most efficient outcome is attained by the solution to the centralized problem,

$$\underset{s}{\text{minimize}} \ \sum_{t=1}^{T} f\left(\delta(t) - \sum_{i=1}^{N} s_i(t)\right) \qquad (6.55)$$

$$\text{subject to } \sum_{t=1}^{T} s_i(t) = 0. \tag{6.56}$$

This is essentially a simplified multi-period optimal power flow with the power balance constraint embedded in the objective. There are an infinite number of optimal centralized solutions, a subset of which are characterized by

$$s_i^c(t) = \gamma_i \left(\delta(t) - \bar{\delta} \right)$$

$$\sum_{i=1}^{N} \gamma_i = 1$$

where the average demand is

$$\bar{\delta} = \frac{1}{T} \sum_{t=1}^{T} \delta(t).$$

In this solution, the storages perfectly flatten the load curve so that the net load served by the generator is $\bar{\delta}$ at all times. The choice of γ determines the fraction of the load curve flattened by each individual storage.

Now suppose that each storage maximizes its own profits and is aware of its effect on the market clearing price, i.e., that

$$\lambda(t) = a \left(\delta(t) - \sum_{i=1}^{N} s_i(t) \right) + b.$$

Storage i then solves

$$\underset{s_i}{\text{maximize}} \sum_{t=1}^{T} \left(a \left(\delta(t) - \sum_{i=1}^{N} s_i(t) \right) + b \right) s_i(t)$$

$$\text{subject to } \sum_{t=1}^{T} s_i(t) = 0,$$

which couples each individual storage's decisions to all of the others'. The resulting collection of N coupled maximizations defines an N-player game. Informally, this game is a type of Cournot competition like in Example 6.2 because the players are using their outputs as means to manipulate prices. Because each player's objective is an equality-constrained concave maximization, we can analytically solve for the PNE:

$$s_i^g(t) = \frac{1}{N+1} \left(\delta(t) - \bar{\delta} \right).$$

At the PNE, the storages again flatten the load but leave some variation so as to preserve market price variations, without which arbitrage yields zero profit.

We quantify the difference between the strategic and ideal centralized outcomes by considering the *efficiency loss*, also known more dramatically as the *price of anarchy* [78]:

$$\Phi = \frac{\sum_{t=1}^{T} f\left(\delta(t) - \sum_{i=1}^{N} s_i^c(t)\right)}{\sum_{t=1}^{T} f\left(\delta(t) - \sum_{i=1}^{N} s_i^g(t)\right)}$$

$$= \frac{T\left(\frac{a}{2}\bar{\delta}^2 + b\bar{\delta}\right)}{\sum_{t=1}^{T} \frac{a}{2}\left(\frac{1}{N+1}\delta(t) + \frac{N}{N+1}\bar{\delta}\right)^2 + b\left(\frac{1}{N+1}\delta(t) + \frac{N}{N+1}\bar{\delta}\right)}.$$

Letting $N \to \infty$, the efficiency loss vanishes, i.e., $\Phi \to 1$.

Example 6.9 illustrates some basic features of markets. A large number of proportionately sized participants generally leads to greater efficiency, which here manifests in the flatness of the net load profile: in the game, variation is allowed to persist so that arbitrage opportunities exist in the market price, as illustrated in Figure 6.5. As the number of participants increases, the influence of each individual storage diminishes, the net load profile flattens, and system efficiency increases.

To a certain extent, this example further validates studying duopolies, as discussed in Section 6.1.3, because they can expose vulnerabilities by capturing the least efficient outcomes. We now give three well-known models, each of which generalizes Example 6.2 or 6.3 and captures a distinctive aspect of power markets.

6.3.1   Supply function equilibrium

In supply function competition, power producers compete through bids stating the amount of power they are willing to provide at each price. Basic results for supply function competition were established in endnote [79] and soon after applied to electricity markets in endnote [80] and then [81], which established the supply function duopoly as a standard approach to market power. Supply functions closely resemble the way power producers declare their costs in real power markets, as described in Section 6.2. Because we only enforce a real power balance, we are essentially considering a strategic

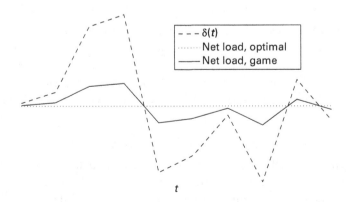

Fig. 6.5   The nominal load without storage, δ, and the net load in the centralized and game outcomes with $N = 3$.

version of Example 6.4, which although extremely crude for pricing is near the tractable boundary of game theory.

We denote each player's supply function by $\sigma_i(\lambda)$, where λ is the market price. It is assumed that demand, $\delta(\lambda)$, is concave and strictly decreasing, and that each player's cost, $f_i(p_i)$, is convex and strictly increasing. In the case that each $\sigma_i(\lambda)$ is a horizontal line for which the market price is entirely determined by demand elasticity, we have Cournot competition of Example 6.2. When they are vertical lines corresponding to any amount of production for a particular price, we recover Bertrand competition of Example 6.3. It is therefore unsurprising that the efficiency of supply function competition is always between the most efficient Bertrand and less efficient Cournot outcomes.

The sum of the supply function must match the elastic demand,

$$\sum_i \sigma_i(\lambda) = \delta(\lambda). \tag{6.57}$$

Each producer attempts to maximize its profits by solving

$$\underset{\sigma_i}{\text{maximize}} \; \lambda \sigma_i(\lambda) - f_i(\sigma_i(\lambda)).$$

Because σ_i is a function defined over a continuum of prices, this is an infinite dimensional optimization problem, which is not amenable to the optimization tools covered in this book. Specializing to a duopoly with symmetric costs, we substitute (6.57) and change the optimization variable to λ to obtain[10]

$$\underset{\lambda}{\text{maximize}} \; \lambda \left(\delta(\lambda) - \sigma_{-i}(\lambda) \right) - f \left(\delta(\lambda) - \sigma_{-i}(\lambda) \right).$$

After differentiating with respect to λ, applying the fact that a symmetric solution must exist, and some further manipulations, we obtain the equilibrium condition

$$\frac{d\sigma(\lambda)}{d\lambda} = \frac{\sigma(\lambda)}{\lambda - \left. \frac{df(p)}{dp} \right|_{p=\sigma(\lambda)}} + \frac{d\delta(\lambda)}{d\lambda},$$

where $\sigma_i(\lambda) = \sigma_{-i}(\lambda) = \sigma(\lambda)$. This is a scalar, nonlinear ordinary differential equation, for which no analytical solution is known to exist in the general case.

The equilibrium condition becomes more manageable upon specializing to linear supply functions of the form $\sigma_i(\lambda) = \alpha_i \lambda$, cf. [82, 83]. Assume further that $f_i(p_i) = c_i p_i^2 / 2$ and that $\delta(\lambda) = \gamma + \beta \lambda$, $\beta < 0$. In this case, we can allow for any number of players with different costs, and the equilibrium condition becomes a system of coupled algebraic equations:

$$\alpha_i = (1 - c_i \alpha_i) \left(-\beta + \sum_{j \neq i} \alpha_j \right). \tag{6.58}$$

[10] We offer somewhat hand-waving justification for this step: suppose $\sigma_i(\lambda)$ is player i's best response to the residual demand, $\delta(\lambda) - \sigma_j(\lambda)$. In choosing $\sigma_i(\lambda)$, it is effectively choosing its real power output, $p_i = \sigma_i(\lambda)$. Assuming monotonicity of $\sigma_i(\lambda)$, this associates a unique value of λ with each p_i, so that maximizing over λ against the residual demand produces the same result. While informal, the approach has nevertheless proven effective in understanding strategic competition in electricity markets.

More recent work has considered supply functions of the form $\sigma_i(\lambda) = \delta - \alpha_i/\lambda$ [84], where δ is an inelastic demand and α_i the player's sole decision parameter. It is shown that when supply functions are restricted to this format and there are more than two players, the efficiency loss as defined in Example 6.9 is tightly bounded above by

$$\Phi = 1 + \frac{1}{N-2},$$

which also approaches unity as $N \to \infty$. As expected, this result also conforms with the wisdom that more market participants means improved competition and efficiency.

6.3.2   Complementarity models

Complementarity models of strategic competition contrast the supply function competition of Section 6.3.1 in that they are computationally scalable and can generically assimilate diverse network details, but they sacrifice modeling of agent costs. Perhaps most significantly, they often assume that elastic demands match the output of power producers as in Cournot competition (Example 6.2). This is nearly the reverse of actual current power system operations, in which production is scheduled to meet mostly inelastic demands.

Given the square matrix M and vector q, the linear complementary problem [85] is to find vectors w and z satisfying

$$w = q + Mz \tag{6.59}$$
$$w, z \geq 0 \tag{6.60}$$
$$w_i z_i = 0. \tag{6.61}$$

Given z, the slack variable w is determined by the first relation. If $M \succ 0$, a unique solution exists and is characterized by the QP

$$\underset{z}{\text{minimize}} \ \ z^T(Mz + q)$$
$$\text{subject to} \ \ Mz + q \geq 0$$
$$z \geq 0.$$

Note that more efficient, specialized solution methods exist, cf. [86, 87].

Now observe that the KKT conditions of Section 6.1.1 for an LP or QP are expressible within the linear complementarity framework by using (6.59) for stationarity and primal feasibility, (6.60) for primal and dual feasibility, and (6.61) for complementary slackness. Moreover, linear complementarity can combine the KKT conditions of multiple optimization problems, even if they have shared variables. This enables us to write the equilibrium conditions for many games with simple structure, guaranteeing the existence and tractability of PNE.

Linear complementarity and generalizations have been widely employed to study games with crude production and relatively detailed transmission modeling, cf. [88–90] and [91]. These approaches, however, can exhibit high sensitivities to seemingly minor modeling details [92, 93].

6.3.3   Capacitated price competition

We now consider a game well suited to strategic competition with hard constraints, for example, from energy storage, generation, or transmission capacity limits. Recall that Bertrand competition predicts that players will undercut each other's prices until their revenues are exactly their production costs, the ideal outcome. Edgeworth [94] augmented Bertrand competition with production capacity limits. In the resulting more realistic model, players engage in undercutting but also increase prices when other players' capacity limits guarantee them a portion of the demand. Of equal significance is that PNE may fail to exist in this *capacitated price* or *Bertrand-Edgeworth competition*.

Whereas supply function models accommodate power producers with complex cost curves, capacitated price competition best describes the simple costs and hard capacity limits of energy storage and has also been used to compare auction formats for power markets with conventional generators [95]. Here we characterize MNE in capacitated price competition and also consider a two-stage game extension where players strategically set capacities prior to the pricing stage, as studied in the seminal work of Kreps and Scheinkman [96].[11]

We present the n-player formulation of endnote [97]. In the pricing stage, each player submits a price, λ_i, and then receives a portion of the demand limited by the player's capacity, c_i. It is assumed that all players share the same linear production costs as in Example 6.3, which are subtracted off. The total demand, d, is allocated price-wise according to the below LP:[12]

$$\underset{p}{\text{minimize}} \quad \lambda_r p_r + \sum_i \lambda_i p_i$$

$$\text{subject to} \quad p_r + \sum_i p_i = d$$

$$0 \leq p_i \leq c_i$$

p_r is the residual demand not captured by the players and is assessed at the exogenous price λ_r. Because p_r has no upper capacity limit, $\lambda_i \leq \lambda_r$ if player i is to receive any share of the demand; for this reason, λ_r is sometimes referred to as the *reservation utility*.

Given the price vector λ, a player's profit is its price times the portion of the demand the player receives:

$$\lambda_i \min \left\{ c_i, \left(d - \sum_{j:\lambda_j < \lambda_i} c_j \right)^+ \right\}.$$

Observe that this game is discontinuous: player i's profits can jump when its price crosses another player's, and consequently there is no guarantee that a PNE exists. An

---

[11] Here it is shown that Bertrand price preceded by capacity competition produces Cournot outcomes in duopolies. We use a simplified model with inelastic demand for which the comparison breaks down.

[12] Demand could be made a more general function of price. Because power demand is nearly inelastic and energy storage often addresses unscheduled power imbalances, it is appropriate in the current context to model demand inelastically.

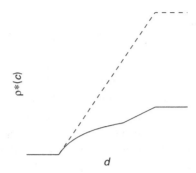

Fig. 6.6   The expected profits of each player, $\rho_i^*(c)$, in a capacitated price duopoly for $\lambda_r = 1$, $c_1 = 1/5$, $c_2 = 3/5$, and $d \in [0, 1]$. The flat regions on the left and right correspond to equilibrium cases one and three, respectively, and the middle region to case two.

MNE, however, does exist [98]. Player i's utility from playing the mixed strategy $m_i(\lambda_i)$ is just the expectation of the above profit function over the mixed strategies:

$$\rho_i(m) = \int_\lambda \left(\prod_j m_j(\lambda_j) \right) \lambda_i \min \left\{ c_i, \left(d - \sum_{j:\lambda_j < \lambda_i} c_j \right)^+ \right\} d\lambda. \tag{6.62}$$

The Nash equilibrium of this game is characterized by the following three cases. Figure 6.6 shows the optimal utilities, $\rho_i^*(c)$, in the duopoly case.

Case 1: $d \leq \sum_{i \neq j} c_i$ for all j
Any subset of $n - 1$ players can accommodate the entire demand. Consequently, there is always a "left-out" player to undercut the others, and the resulting PNE is $\lambda_i = 0$ and $\rho_i^*(c) = 0$. This is the standard Bertrand competition outcome.

Case 2: $\sum_{i \neq j} c_i < d < \sum_i c_i$ for some j
An MNE exists in which player i's profit is

$$\rho_i^*(c) = \frac{\lambda_r \min\{c_i, d\}}{\min\{c_{max}, d\}} \left(d + c_{max} - \sum_j c_j \right),$$

where $c_{max} = \max_i c_i$. While seemingly narrow relative to the ranges of scenarios corresponding to the other cases, this intermediary case is often the most realistic because the total capacity is neither under- nor oversized for the demand.

Case 3: $d \geq \sum_i c_i$
All players receive their full capacity, c_i, regardless of their price. Because there is no incentive to compete for the demand, the PNE is $\lambda_i = \lambda_r$, and each player's profit is $\rho_i^*(c) = \lambda_r c_i$.

This equilibrium tells us a number of interesting things about capacitated price competition. The first and third cases respectively describe over- and undersized capacity and produce outcomes one would intuitively expect. As shown in Figure 6.6, the expected profits in the second case interpolate between the other two cases. More interestingly, no PNE exists in this case, indicating that players will vary their actions to gain competitive advantage, much in the way that retail stores frequently hold reduced-price sale events.

Example 6.10 *Local market power due to transmission congestion.* It has long been observed that local producers can gain market power at nodes whose import capabilities are limited by transmission constraints, cf. [99, 100]. This example uses capacitated price competition to analyze this phenomenon.

Suppose two nodes are connected by a lossless transmission line of capacity $\bar{s}$. The nodes have unlimited power generation p_i and identical demands d, as shown in Figure 6.7. The loads will not purchase power at prices above the reservation utility, λ_r.

The producer at each node sets a price, $\lambda_i \leq \lambda_r$, and then profits $\lambda_i p_i$. Producer i's profits are given by

$$\lambda_i \begin{cases} d + \min\{d, \bar{s}\} & \text{if} \quad \lambda_i < \lambda_{-i} \\ \max\{d - \bar{s}, 0\} & \text{if} \quad \lambda_i > \lambda_{-i} \\ d & \text{if} \quad \lambda_i = \lambda_{-i} \end{cases}.$$

Note that this is different from the nodal pricing mechanism of Section 6.2.1.

Observe that this scenario is mathematically equivalent to capacitated price competition between two identical firms with capacities $d + \bar{s}$ and total demand $2d$. We may thus transcribe the capacitated price competition equilibria, which are listed case-wise below:

- $d \leq \bar{s}$: Each producer can fulfill the other node's demand, leading them to undercut each other to the PNE $\lambda_1 = \lambda_2 = 0$, as in standard Bertrand competition.
- $0 < \bar{s} < d$: No PNE exist. At the MNE, each producer randomizes its prices according to the distribution

$$\mu(\lambda_i) = \frac{\lambda_r(d - \bar{s})}{2\bar{s}\lambda_i^2}, \quad \lambda_i \in \left[\frac{\lambda_r(d - \bar{s})}{d + \bar{s}}, \lambda_r\right],$$

and has expected profits $\lambda_r(d - \bar{s})$. Observe that as $\bar{s}$ approaches zero, the support of the mixed strategy approaches the point λ_r, and the MNE collapses to the PNE in the second case. As $\bar{s}$ approaches d, the lower limit of the support approaches zero, and the probabilities of all prices except $\lambda_i = 0$ approach zero.
- $\bar{s} = 0$: At the PNE, both producers charge the reservation utility, λ_r, because they face no competition from each other.

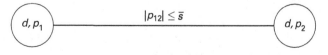

Fig. 6.7   Market power in a congested two-node transmission network.

We see that, by constraining supply, transmission congestion guarantees producers a portion of the demand regardless of their prices, enabling them to profit above their marginal costs. In other words, congestion brings each producer closer to a local monopoly.

Suppose now that capacities are determined at a prior stage and that player i's capacity cost is $\gamma_i c_i$. This could represent the fixed costs of construction-planning decisions or opportunity and standby operation costs of capacity commitments. Player i then maximizes its combined profits

$$\rho_i^*(c) - \gamma_i c_i,$$

which defines the capacity game. A continuous range of PNE exists, each of which satisfies

$$\sum_i c_i = d$$

and

$$\frac{\lambda_r - \gamma_i}{2\lambda_r - \gamma_i}(c_i + c_{\max}) \le c_i \le c_{\max}$$

for all non-maximal i. At the PNE, the player community sizes its aggregate capacity to exactly match demand at equilibrium, so that each player's profit is $(\lambda_r - \gamma_i)c_i$. From a social perspective, this is the least competitive, maximally inefficient outcome.

Let us now discuss the relevance of the above model, which is somewhat less established for modeling electricity markets than the preceding two. The capacitated price description handles hard capacity constraints well, like with transmission congestion in Example 6.10 and, in particular, energy storage, which has both hard capacity limits and near-uniform marginal costs. However, capacitated price competition as shown above is deficient for storage in three respects. First, it is a *single-shot* game, which cannot account for the fundamentally dynamic nature of energy storage, cf. Section 4.1.2. Second, one of the primary purposes of energy storage is to absorb unscheduled, random power imbalances, which are deterministic in this model. Finally, each player receives its distinct bid price in the assumed allocation mechanism, as in a *divisible, discriminatory price auction* [101]. While baseload power markets like those in Section 6.2 are equatable with uniform price auctions, it has been frequently debated whether reserve markets should be run as uniform or discriminatory auctions, cf. [102]. Taken together, capacitated price competition does capture some essential aspects of competition under hard constraints and is a promising substrate for more detailed models.

6.4   Summary

Electricity prices are obtained from the Lagrangian dual of optimal power flow, which is called economic dispatch in this context. Dual prices only support economic dispatch if it is convex and thus has strong duality, necessitating the use of the convex power flow approximations of Chapter 3. A key concept is that details absent from an economic dispatch routine will not be reflected by the resulting nodal or transmission pricing mechanism. Consequently, relaxations of power flow can straightforwardly price many features absent from the basic linearized model, in particular losses and reactive power.

We have also seen that the mathematical assumptions underpinning electricity markets virtually never hold in practice. One reason is that even though prices may be derived from convex economic dispatch, resources must first be scheduled via unit commitment routines (Section 4.3), which have nonconvex integer constraints. The situation is partially addressed in Section 6.2.4, but no perfect remedy is known to exist.

Another major reason electricity markets do not function ideally is that many market participants are not price-takers. Large agents can often perceive the effect of their choices on market mechanisms like nodal prices, leading them to deviate from socially optimality to increase their individual profits. Alternatively, producers may intentionally inflate their supply bids to get paid more than their costs; there are many ways for electricity market participants to gain market power. Game theory can explain some such behaviors, but its inherent intractability makes it difficult to use with detailed power flow and production cost models. Compared to conventional optimization problems, market power is extremely difficult to address, because we must use a more general and less tractable framework to model a more diverse range of scenarios.

The deficiencies of electricity markets open many avenues for future work. Basic market architectures must change to accommodate renewable power sources and active demand-side participation. Game theory is an active area of theoretical research, especially with the advents of algorithmic game theory and mechanism design, which we briefly discuss in Section 7.3.2. Decades after the initial steps toward deregulation, entirely new players are entering the power industry like solar panel installers and demand response aggregators, each bringing new, coupled engineering and economic challenges.

Problems

6.1 In Section 6.2.1, the shadow price associated with line congestion, χ_{ij}, is sometimes but not always the difference of the nodal prices, $\lambda_j - \lambda_i$. Construct a three-node network in which this is true and another for which it is not.

6.2 Show that, for the linearized model in radial networks, the difference of the nodal prices is equal to the intermediary line's shadow price.

6.3 Consider nodal pricing with the linearized model in Section 6.2.1, and suppose there are no nodal power limits. Recall the definition of a competitive equilibrium (from Section 6.1.2) as a vector of powers p, angles θ, and prices λ satisfying (6.13)-(6.15) and (6.16). Give an example of a competitive equilibrium that is not an economic dispatch.

6.4 Derive the Lagrangian dual of economic dispatch with the objective $\sum_i a_i p_i^2 + b_i p_i$ and

(a)   the linearized power flow (Feasible Set 3.3);
(b)   the SOC branch flow relaxation (Feasible Set 3.13);
(c)   the SD voltage-polar coordinate power flow relaxation (Feasible Set 3.11).

6.5 Derive congestion and loss FGRs (flowgate rights) as described in Section 6.2.3 for the SD relaxation in the previous exercise.

6.6 Derive the equilibrium for the Bertrand duopoly of Example 6.3 when each player has a different marginal cost, i.e., $c_i \neq c_{-i}$.

6.7 Consider a duopoly version of the capacitated pricing competition of Section 6.3.3. Suppose that instead of each player being paid its bid price, λ_i, players are paid the highest accepted bid price, so that player i's payoff is

$$
\begin{cases}
\lambda_i d & \text{if} \quad \lambda_i < \lambda_{-i}, \ d \leq c_i \\
\lambda_{-i} c_i & \text{if} \quad \lambda_i < \lambda_{-i}, \ d > c_i \\
\lambda_i c_i \min\left\{\frac{d}{c_i + c_{-i}}, 1\right\} & \text{if} \quad \lambda_i = \lambda_{-i} \\
\lambda_i \min\left\{(d - c_{-i})^+, c_i\right\} & \text{if} \quad \lambda_i > \lambda_{-i}.
\end{cases}
$$

Derive the price equilibrium for this game.

6.8 Consider the setup of Section 6.3.3, and suppose that the pricing game is preceded by a capacity setting game and that each player has a maximum capacity limit $c_i \leq \bar{c}_i$. Derive the capacity PNE for this game.

References

[1] P. L. Joskow and R. Schmalensee, *Markets for Power: An Analysis of Electrical Utility Deregulation*. MIT Press, Mar. 1988.

[2] P. L. Joskow, "Restructuring, competition and regulatory reform in the U.S. electricity sector," *Journal of Economic Perspectives*, vol. 11, no. 3, pp. 119–138, 1997.

[3] ——, "California's electricity crisis," *Oxford Review of Economic Policy*, vol. 17, no. 3, pp. 365–388, 2001.

[4] P. Joskow, "Lessons learned from electricity market liberalization," *Energy Journal*, vol. 29, no. 2, pp. 9–42, 2008.

[5] R. Munson, *From Edison to Enron: The Business of Power and What It Means for the Future of Electricity*. Praeger Publishers, 2005.

[6] S. Stoft, *Power System Economics: Designing Markets for Electricity*. Wiley-IEEE Press, 2002.

[7] M. Shahidehpour, H. Yamin, and Z. Li, *Market Operations in Electric Power Systems (Forecasting, Scheduling, and Risk Management)*. Wiley-IEEE Press, 2002.

[8] D. Kirschen and G. Strbac, *Fundamentals of Power System Economics*. Wiley, 2004.

[9] R. E. Bohn, M. C. Caramanis, and F. C. Schweppe, "Optimal pricing in electrical networks over space and time," *RAND Journal of Economics*, vol. 15, no. 3, pp. 360–376, 1984.

[10] F. C. Schweppe, M. C. Caramanis, R. D. Tabors, and R. E. Bohn, *Spot Pricing of Electricity*. Boston: Kluwer Academic Publishers, 1988.

[11] F. P. Kelly, A. K. Maulloo, and D. K. H. Tan, "Rate control for communication networks: Shadow prices, proportional fairness and stability," *Journal of the Operational Research Society*, vol. 49, no. 3, pp. 237–252, 1998.

[12] S. H. Low and D. E. Lapsley, "Optimization flow control - I: Basic algorithm and convergence," *IEEE/ACM Transactions on Networking*, vol. 7, no. 6, pp. 861–874, Dec. 1999.

[13] R. Srikant, *The Mathematics of Internet Congestion Control*, ser. Systems & Control. Birkhäuser, 2004.

[14] R. T. Rockafellar, *Convex Analysis*, ser. Princeton Landmarks in Mathematics and Physics. Princeton University Press, 1996.

[15] D. Bertsekas, A. Nedić, and A. Ozdaglar, *Convex Analysis and Optimization*, ser. Athena Scientific Optimization and Computation Series. Athena Scientific, 2003.

[16] J. Nocedal and S. J. Wright, *Numerical Optimization*. Springer, 2000.

[17] D. P. Bertsekas, *Nonlinear Programming*. Athena Scientific, 1999.

[18] S. Boyd and L. Vandenberghe, *Convex Optimization*. New York: Cambridge University Press, 2004.

[19] K. Arrow and F. Hahn, *General Competitive Analysis*. Holden-Day, 1971.

[20] H. Varian, *Microeconomic Analysis*. Norton, 1978.

[21] A. Mas-Colell, M. Whinston, and J. Green, *Microeconomic Theory*. Oxford University Press, 1995.

[22] J. Lavaei and S. Low, "Zero duality gap in optimal power flow problem," *IEEE Transactions on Power Systems*, vol. 27, no. 1, pp. 92–107, Feb. 2012.

[23] F. Wu, P. Varaiya, P. Spiller, and S. Oren, "Folk theorems on transmission access: Proofs and counterexamples," *Journal of Regulatory Economics*, vol. 10, pp. 5–23, 1996.

[24] H. A. Simon, "A behavioral model of rational choice," *Quarterly Journal of Economics*, vol. 69, no. 1, pp. 99–118, 1955.

[25] D. Fudenberg and J. Tirole, *Game Theory*. MIT Press, 1991.

[26] M. Osborne and A. Rubinstein, *A Course in Game Theory*. MIT Press, 1994.

[27] G. Debreu, "A social equilibrium existence theorem," *Proceedings of the National Academy of Sciences of the United States of America*, vol. 38, no. 10, pp. 886–893, Oct. 1952.

[28] I. L. Glicksberg, "A further generalization of the Kakutani fixed point theorem, with application to Nash equilibrium points," *Proceedings of the American Mathematical Society*, vol. 3, no. 1, pp. 170–174, 1952.

[29] K. Fan, "Fixed-point and minimax theorems in locally convex topological linear spaces," *Proceedings of the National Academy of Sciences of the United States of America*, vol. 38, no. 2, pp. 121–126, 1952.

[30] J. B. Rosen, "Existence and uniqueness of equilibrium points for concave n-person games," *Econometrica*, vol. 33, no. 3, pp. 520–534, 1965.

[31] J. F. Nash, "Equilibrium points in n-person games," *Proceedings of the National Academy of Sciences*, vol. 36, no. 1, pp. 48–49, 1950.

[32] J. F. Nash Jr., "The bargaining problem," *Econometrica*, pp. 155–162, 1950.

[33] ——, "Non-cooperative games," *The Annals of Mathematics*, vol. 54, no. 2, pp. 286–295, Sept. 1951.

[34] A.-A. Cournot, *Recherches sur les principes mathématiques de la théorie des richesses.* chez L. Hachette, 1838.

[35] J. Bertrand, "Theorie mathematique de la richesse sociale," *Journaldes Savants*, pp. 499–508, 1883.

[36] C. Daskalakis, P. Goldberg, and C. Papadimitriou, "The complexity of computing a Nash equilibrium," *SIAM Journal on Computing*, vol. 39, no. 1, pp. 195–259, 2009.

[37] N. Nisan, T. Roughgarden, E. Tardos, and V. V. Vazirani, *Algorithmic Game Theory*. New York: Cambridge University Press, 2007.

[38] J. Von Neumann and O. Morgenstern, *Theory of Games and Economic Behavior*. Princeton University Press, 1944.

[39] G. Andersson, P. Donalek, R. Farmer, N. Hatziargyriou, I. Kamwa, P. Kundur, N. Martins, J. Paserba, P. Pourbeik, J. Sanchez-Gasca, R. Schulz, A. Stankovic, C. Taylor, and V. Vittal, "Causes of the 2003 major grid blackouts in North America and Europe, and recommended means to improve system dynamic performance," *IEEE Transactions on Power Systems*, vol. 20, no. 4, pp. 1922–1928, Nov. 2005.

[40] P. Pourbeik, P. Kundur, and C. Taylor, "The anatomy of a power grid blackout - root causes and dynamics of recent major blackouts," *IEEE Power and Energy Magazine*, vol. 4, no. 5, pp. 22–29, Oct. 2006.

[41] J. Green and J. Laffont, *Incentives in Public Decision-Making*, ser. Studies in Public Economics. North-Holland Pub. Co., 1979.

[42] N.-H. von der Fehr and D. Harbord, "Spot market competition in the UK electricity industry," *Economic Journal*, vol. 103, no. 418, pp. 531–546, May 1993.

[43] D. Kirschen, "Demand-side view of electricity markets," *IEEE Transactions on Power Systems*, vol. 18, no. 2, pp. 520–527, 2003.

[44] M. Baughman, S. Siddiqi, and J. Zarnikau, "Advanced pricing in electrical systems. I. Theory," *IEEE Transactions on Power Systems*, vol. 12, no. 1, pp. 489–495, Feb. 1997.

[45] ——, "Advanced pricing in electrical systems. II. Implications," *IEEE Transactions on Power Systems*, vol. 12, no. 1, pp. 496–502, 1997.

[46] T. Wu, M. Rothleder, Z. Alaywan, and A. Papalexopoulos, "Pricing energy and ancillary services in integrated market systems by an optimal power flow," *IEEE Transactions on Power Systems*, vol. 19, no. 1, pp. 339–347, 2004.

[47] J. Arroyo and F. Galiana, "Energy and reserve pricing in security and network-constrained electricity markets," *IEEE Transactions on Power Systems*, vol. 20, no. 2, pp. 634–643, 2005.

[48] F. Galiana, F. Bouffard, J. Arroyo, and J. Restrepo, "Scheduling and pricing of coupled energy and primary, secondary, and tertiary reserves," *Proceedings of the IEEE*, vol. 93, no. 11, pp. 1970–1983, 2005.

[49] E. Kahn and R. Baldick, "Reactive power is a cheap constraint," *The Energy Journal*, vol. 15, pp. 191–201, 1994.

[50] S. Hao and A. Papalexopoulos, "Reactive power pricing and management," *IEEE Transactions on Power Systems*, vol. 12, no. 1, pp. 95–104, Feb. 1997.

[51] J. Lamont and J. Fu, "Cost analysis of reactive power support," *IEEE Transactions on Power Systems*, vol. 14, no. 3, pp. 890–898, Aug. 1999.

[52] J. Barquin Gil, T. San Roman, J. Alba Rios, and P. Sanchez Martin, "Reactive power pricing: IEEE Transactions on Power Systems conceptual framework for remuneration and charging procedures," *IEEE Transactions on Power Systems*, vol. 15, no. 2, pp. 483–489, May 2000.

[53] J. Zhong and K. Bhattacharya, "Toward a competitive market for reactive power," *IEEE Transactions on Power Systems*, vol. 17, no. 4, pp. 1206–1215, Nov. 2002.

[54] S. Borenstein, M. Jaske, and A. Rosenfeld, "Dynamic pricing, advanced metering, and demand response in electricity markets," UC Berkeley: Center for the Study of Energy Markets, Tech. Rep., 2002.

[55] S. Borenstein, "The long-run efficiency of real-time electricity pricing," *Energy Journal*, vol. 26, no. 3, pp. 93–116, 2005.

[56] A. Faruqui and S. Sergici, "Household response to dynamic pricing of electricity: A survey of 15 experiments," *Journal of Regulatory Economics*, vol. 38, no. 2, pp. 193–225, 2010.

[57] M. Roozbehani, M. A. Dahleh, and S. K. Mitter, "Volatility of power grids under real-time pricing," *IEEE Transactions on Power Systems*, vol. 27, no. 4, pp. 1926–1940, Nov. 2012.

[58] M. Caramanis and J. Foster, "Uniform and complex bids for demand response and wind generation scheduling in multi-period linked transmission and distribution markets," in *IEEE Conference on Decision and Control and European Control Conference*, 2011, pp. 4340–4347.

[59] J. Warrington, P. Goulart, S. Mariethoz, and M. Morari, "A market mechanism for solving multi-period optimal power flow exactly on AC networks with mixed participants," in *American Control Conference*, June 2012, pp. 3101–3107.

[60] J. Tsitsiklis and Y. Xu, "Pricing of fluctuations in electricity markets," in *IEEE 51st Annual Conference on Decision and Control*, Dec. 2012, pp. 457–464.

[61] FERC, "Order no. 755: Frequency regulation compensation in the organized wholesale power markets," Oct. 2011.

[62] A. Berger and F. Schweppe, "Real time pricing to assist in load frequency control," *IEEE Transactions on Power Systems*, vol. 4, no. 3, pp. 920–926, Aug. 1989.

[63] J. A. Taylor, A. Nayyar, D. S. Callaway, and K. Poolla, "Consolidated dynamic pricing of power system regulation," *IEEE Transactions on Power Systems*, vol. 28, no. 4, pp. 4692–4700, 2013.

[64] J. Rosellón and T. Kristiansen, Eds., *Financial Transmission Rights: Analysis, Experiences and Prospects*. Springer, 2013.

[65] W. W. Hogan, "Contract networks for electric power transmission," *Journal of Regulatory Economics*, vol. 4, pp. 211–242, 1992.

[66] H.-P. Chao, S. Peck, S. Oren, and R. Wilson, "Flow-based transmission rights and congestion management," *Electricity Journal*, vol. 13, no. 8, pp. 38–58, 2000.

[67] R. O'Neill, U. Helman, B. Hobbs, W. R. Stewart Jr., and M. Rothkopf, "A joint energy and transmission rights auction: Proposal and properties," *IEEE Transactions on Power Systems*, vol. 17, no. 4, pp. 1058–1067, 2002.

[68] W. Hogan, "Financial transmission right formulations," *Report, Center for Business and Government, John F. Kennedy School of Government, Harvard University, Cambridge, MA*, 2002.

[69] J. Bialek, "Tracing the flow of electricity," *Generation, Transmission and Distribution, IEE Proceedings*, vol. 143, no. 4, pp. 313–320, July 1996.

[70] D. Kirschen, R. Allan, and G. Strbac, "Contributions of individual generators to loads and flows," *IEEE Transactions on Power Systems*, vol. 12, no. 1, pp. 52–60, 1997.

[71] P. Kattuman, R. Green, and J. Bialek, "Allocating electricity transmission costs through tracing: A game-theoretic rationale," *Operations Research Letters*, vol. 32, no. 2, pp. 114–120, 2004.

[72] J. Bialek and P. A. Kattuman, "Proportional sharing assumption in tracing methodology," *Generation, Transmission and Distribution, IEE Proceedings*, vol. 151, no. 4, pp. 526–532, 2004.

[73] R. P. O'Neill, P. M. Sotkiewicz, B. F. Hobbs, M. H. Rothkopf, and W. R. Stewart Jr., "Efficient market-clearing prices in markets with nonconvexities," *European Journal of Operational Research*, vol. 164, no. 1, pp. 269–285, 2005.

[74] W. W. Hogan and B. J. Ring, "On minimum-uplift pricing for electricity markets," *Electricity Policy Group*, 2003.

[75] P. R. Gribik, W. W. Hogan, and S. L. Pope, "Market-clearing electricity prices and energy uplift," *Electricity Policy Group*, 2007.

[76] G. Wang, U. Shanbhag, T. Zheng, E. Litvinov, and S. Meyn, "An extreme-point subdifferential method for convex hull pricing in energy and reserve markets – Part I: Algorithm structure," *IEEE Transactions on Power Systems*, vol. 28, no. 3, pp. 2111–2120, 2013.

[77] ——, "An extreme-point subdifferential method for convex hull pricing in energy and reserve markets – Part II: Algorithm structure," *IEEE Transactions on Power Systems*, vol. 28, no. 3, pp. 2111–2120, 2013.

[78] T. Roughgarden, *Selfish Routing and the Price of Anarchy*. The MIT Press, 2005.

[79] P. D. Klemperer and M. A. Meyer, "Supply function equilibria in oligopoly under uncertainty," *Econometrica*, vol. 57, no. 6, pp. 1243–1277, 1989.

[80] F. Bolle, "Supply function equilibria and the danger of tacit collusion: The case of spot markets for electricity," *Energy Economics*, vol. 14, no. 2, pp. 94–102, 1992.

[81] R. J. Green and D. M. Newbery, "Competition in the British electricity spot market," *Journal of Political Economy*, vol. 100, no. 5, pp. 929–953, 1992.

[82] R. Green, "Increasing competition in the British electricity spot market," *Journal of Industrial Economics*, vol. 44, no. 2, pp. 205–216, 1996.

[83] R. Baldick, R. Grant, and E. Kahn, "Theory and application of linear supply function equilibrium in electricity markets," *Journal of Regulatory Economics*, vol. 25, pp. 143–167, 2004.

[84] R. Johari and J. N. Tsitsiklis, "Parameterized supply function bidding: Equilibrium and efficiency," *Operations Research*, vol. 59, no. 5, pp. 1079–1089, 2011.

[85] R. W. Cottle and G. B. Dantzig, "Complementary pivot theory of mathematical programming," *Linear Algebra and Its Applications*, vol. 1, no. 1, pp. 103–125, 1968.

[86] R. W. Cottle and Y.-Y. Chang, "Least-index resolution of degeneracy in linear complementarity problems with sufficient matrices," *SIAM Journal on Matrix Analysis and Applications*, vol. 13, no. 4, pp. 1131–1141, 1992.

[87] R. Cottle, J. Pang, and R. Stone, *The Linear Complementarity Problem*, ser. Classics in Applied Mathematics. Society for Industrial and Applied Mathematics, 1992.

[88] J. B. Cardell, C. C. Hitt, and W. W. Hogan, "Market power and strategic interaction in electricity networks," *Resource and Energy Economics*, vol. 19, pp. 109–137, 1997.

[89] B. Hobbs, C. Metzler, and J.-S. Pang, "Strategic gaming analysis for electric power systems: An MPEC approach," *IEEE Transactions on Power Systems*, vol. 15, no. 2, pp. 638–645, May 2000.

[90] B. Hobbs, "Linear complementarity models of Nash-Cournot competition in bilateral and POOLCO power markets," *IEEE Transactions on Power Systems*, vol. 16, no. 2, pp. 194–202, May 2001.

[91] S. Gabriel, A. Conejo, J. Fuller, B. Hobbs, and C. Ruiz, *Complementarity Modeling in Energy Markets*, ser. International Series in Operations Research & Management Science. Springer, 2013.

[92] K. Neuhoff, J. Barquin, M. G. Boots, A. Ehrenmann, B. F. Hobbs, F. A. Rijkers, and M. Vázquez, "Network-constrained Cournot models of liberalized electricity markets: The devil is in the details," *Energy Economics*, vol. 27, no. 3, pp. 495–525, 2005.

[93] B. Willems, I. Rumiantseva, and H. Weigt, "Cournot versus supply functions: What does the data tell us?" *Energy Economics*, vol. 31, no. 1, pp. 38–47, 2009.

[94] F. Edgeworth, "The pure theory of monopoly," in *Papers Relating to Political Economy*. Macmillan and Co., Ltd., 1925, vol. 1, pp. 111–142.

[95] N. Fabra, N.-H. von der Fehr, and D. Harbord, "Designing electricity auctions," *The RAND Journal of Economics*, vol. 37, no. 1, pp. 23–46, 2006.

[96] D. M. Kreps and J. A. Scheinkman, "Quantity precommitment and Bertrand competition yield Cournot outcomes," *The Bell Journal of Economics*, vol. 14, no. 2, pp. 326–337, 1983.

[97] D. Acemoglu, K. Bimpikis, and A. Ozdaglar, "Price and capacity competition," *Games and Economic Behavior*, vol. 66, no. 1, pp. 1–26, 2009.

[98] D. Acemoglu and A. Ozdaglar, "Competition and efficiency in congested markets," *Math. Oper. Res.*, vol. 32, pp. 1–31, Feb. 2007.

[99] P. L. Joskow and J. Tirole, "Transmission rights and market power on electric power networks," *The RAND Journal of Economics*, vol. 31, no. 3, pp. 450–487, 2000.

[100] S. Borenstein, J. Bushnell, and S. Stoft, "The competitive effects of transmission capacity in a deregulated electricity industry," *The RAND Journal of Economics*, vol. 31, no. 2, pp. 294–325, 2000.

[101] V. Krishna, *Auction Theory*. Academic Press/Elsevier, 2009.

[102] N. Fabra, N.-H. von der Fehr, and D. Harbord, "Modeling electricity auctions," *Electricity Journal*, vol. 15, no. 7, pp. 72–81, 2002.

7 Future directions

This final chapter briefly surveys some challenging and newer topics that have not crystallized to the same extent as the material in previous chapters, but the further development of which could render great benefits to power systems.

7.1 Uncertainty modeling

Throughout this book we have studied deterministic problems with only a few exceptions such as inventory control in Section 4.1.2 and linear quadratic regulation in Section 4.2.2. Uncertainty, however, is pervasive in power systems and can be modeled with a handful of mechanistic frameworks. The three main physical sources of uncertainty are loads, failures, and intermittent renewable energy sources like wind and solar. Because load uncertainty is mostly the result of numerous smaller random events, it is often modeled using Gaussian random variables, while failures are discrete events and must thus be modeled using discrete random variables, cf. [1]. Renewable uncertainty depends on the energy source; for example, wind speed is described by the Weibull distribution, and the distribution of wind power is the derived distribution obtained from cubing wind speed [2]. In addition to physical sources of uncertainty, many participants in electricity markets face financial uncertainty from nodal electricity prices.

7.1.1 Stochastic programming

In stochastic programming, uncertainty is modeled using probability distributions [3, 4]. A standard formulation in stochastic programming is the *two-stage problem with recourse*:

$$\underset{x}{\text{minimize}} \ f(x) + \underset{\delta}{\mathbb{E}} \ [h(x, \delta)]$$
$$\text{subject to} \ g_i(x) \leq 0.$$

The random variable δ represents information that is unknown at the time x is chosen. The function $h(x, \delta)$ is the optimal objective of another optimization problem parameterized by x and δ, given by

$$h(x, \delta) \ = \ \underset{y}{\min} \ k(y, x, \delta)$$
$$\text{subject to} \ l_i(y, x, \delta) \leq 0.$$

The optimization over y is often referred to as a recourse decision because it is made after the realization of x and δ. One can envision even more general formulations in which information is revealed sequentially over many times stages. Such formulations can lead to a number of intractable problems, but some special cases are efficiently solvable, for example when all objective and constraint functions are linear. Multi-stage models with recourse have found wide application in power systems, particularly for scheduling reserves in unit commitment routines [5, 6].

Another popular flavor of stochastic programming is the *chance constraint*:

$$\text{Prob}(f(x, \delta) \leq 0) \geq \alpha. \tag{7.1}$$

Now we are requiring the probability of satisfying an inequality that depends on the decision variable x and random variable δ to be greater than $\alpha \in (0, 1)$. Chance constraints are usually nonconvex and hence computationally troublesome. Moreover, the parameter α, which would usually be close to one for most engineering problems, is often somewhat arbitrary and yet can exert dramatic influence; for example, guaranteeing adequate power generation 99.9% of the time may cost orders of magnitude more than 99% of the time.

In the special case that $f(x, \delta) = \delta^T x - \gamma$ and δ is Gaussian with mean μ and covariance Σ, (7.1) can be written as the SOC constraint

$$\mu^T x + \Phi^{-1}(\alpha) \left\| \sqrt{\Sigma} x \right\| \leq \gamma,$$

where Φ^{-1} is inverse Gaussian cumulative distribution function [7]. This constraint has been used to optimize supply function bids in electricity markets when the nodal price of electricity is uncertain [8].

7.1.2 Robust optimization

Robust optimization is a computationally cheaper way of incorporating uncertainty than stochastic programming [9–11]. Suppose that δ is uncertain. Whereas stochastic programming attributes a probability distribution to δ, robust optimization says that δ must be allowed to take on any value in a prescribed convex set. For example, we may want to solve the optimization problem

$$\begin{aligned} \underset{x}{\text{minimize}} \quad & f(x) \\ \text{subject to} \quad & g(x, \delta) \leq 0 \end{aligned}$$

but are uncertain of δ. The *robust counterpart* is then a *semi-infinite* optimization of the form

$$\begin{aligned} \underset{x}{\text{minimize}} \quad & f(x) \\ \text{subject to} \quad & g(x, \delta) \leq 0 \quad \forall \delta \in \Delta, \end{aligned}$$

where Δ is the predefined set of possible realizations of δ. The strength of robust optimization is that the robust counterpart may be only slightly more difficult than the case of no uncertainty. For example, in endnote [11], the robust counterpart of an LP is an

SOCP if Δ above can be represented by SOC constraints. Robust optimization traces its roots to robust control [12, 13], which has been applied widely in power system control, cf. [14, 15]. As would be expected, robust optimization has found similar application to stochastic programming, including price uncertainty in load control [16], unit commitment [17], and reserve scheduling [18].

7.2   Decentralization and distributed optimization

There are many settings where it is useful to decentralize, for example to solve a problem on multiple processors or because no central authority exists [19]. A typical decentralized control problem could be to guarantee the stability of the system

$$\dot{x}_i = f_i(x_i, u_i) + \sum_{j \neq i} f_{ij}(x_i, x_j, u_i, u_j),$$

$$y_i = g_i(x_i)$$

where x_i is a state vector representing subsystem i, and u_i is its control input. Although the system is coupled to other subsystems through f_{ij}, its observation, y_i, only contains local information about x_i and perhaps limited information about other agents' states.

In power systems, the motivation for decentralization has often arisen from its geographically distributed nature and fast time scales, such as in Section 4.2. In other words, parts of the system are physically too fast and distant to consult with a central authority in their decision processes. Consequently, many aspects of the power system are already decentralized; for example, *speed governors* providing *automatic generation control*, which locally maintain frequency on fast time scales. On the other hand, slower actions like optimal power flow-based generator dispatch are dictated from system operators every few minutes. In the spectrum between speed governors and optimal power flow, there are numerous formulations of decentralized, real-time controllers that achieve system-level objectives with minimal or no communication; see endnotes [20–23] for recent approaches.

Distributed optimization is an overlapping topic in which a single optimization problem is solved piecemeal by multiple agents who can iteratively exchange information [24]. One should note the conceptual similarity to our use of dual multiplier prices in Section 6.2 to induce selfish market participants to independently realize socially good outcomes; indeed, many distributed optimization algorithms are based on duality. A candidate for distributed optimization could look like

$$\underset{x}{\text{minimize}} \quad \sum_i f_i(x_i)$$

$$\text{subject to} \quad Ax = b$$

$$x \geq 0.$$

This is a standard form LP with a separable objective. A plethora of techniques are available if each objective term is convex. Each agent i would then minimize a local objective typically composed of $f_i(x_i)$, the constraint (e.g., via the Lagrangian), and updated information from other agents. Example 7.1 gives a distributed algorithm for a simplistic instance of the above scenario.

Example 7.1 *Distributed optimal power flow with quadratic costs and the power balance constraint.* Recall the economic dispatch problem in Example 6.4:

$$\underset{p}{\text{minimize}} \ \sum_i a_i p_i^2 + b_i p_i$$

$$\text{subject to} \ \sum_i p_i = 0.$$

The analytical solution to this problem is given by

$$\lambda = \frac{\sum_i b_i/a_i}{\sum_i 1/a_i}$$

$$p_j = \frac{\lambda - b_j}{2a_j},$$

where the dual multiplier λ is interpretable as the market clearing price. Here, we construct an iterative distributed algorithm for this problem.

Suppose that there are n nodes in the network, and each one can send messages directly to a limited number of others. We further assume that this communication can be either one- or two-way, i.e., that some communication links can only carry information from one node to another and not vice versa.

We model the communication network as a graph with the same nodes as the power system but with arcs that can be directed or undirected (see Section 2.3.2). (In this case, the weight of arc ij is $w_{ij} = 1$ if i can send messages to j and $w_{ij} = 0$ otherwise.) We assume that the graph is connected, i.e., that a directed path exists from every node to every other node. We let σ_{ij} denote the *shortest path* from i to j through the communication graph, and let $m = \max_{ij} |\sigma_{ij}|$ denote the length of the longest shortest path. This implies that any node can communicate with any other node via a maximum of $m \leq n - 1$ messages, where we have assumed that messages between non-adjacent nodes are relayed by intermediary nodes along the connecting path. We assume somewhat unrealistically that each node knows the full graph structure, i.e., knows the shortest path from itself to each other node.

The following is then a distributed algorithm for the above optimal power flow problem. Each node i simply transmits its parameters, a_i and b_i, to all other nodes through the shortest paths, which we assume are known to all. Because the longest shortest path is m arcs long, each node will know a_i and b_i for all other nodes after a maximum of m communication iterations. At this point, each node can compute λ and p_i above on its own and hence implement the optimal solution.

Distributed optimization is relevant to a number of modern scenarios with small resources like distributed generators, energy storage, electric vehicles, and active loads, many of which are embedded in low-voltage distribution systems. These resources' communication capabilities are limited by their large numbers coupled with the fact that their power hardware does not dwarf the cost of communication hardware. The long communication links necessary for direct centralized operation may therefore be too expensive, making *neighbor-to-neighbor* communications a more desirable but less efficient alternative. Finally, these resources often operate on slower, less critical time scales than transmission-level generators, leaving more time for iterated exchanges of information and convergence to better solutions, like in the above example.

Beyond the centralized problem being simple enough to solve analytically, the above example was an easy distributed optimization problem because each node knew the full graph structure, which is not usually assumed in the literature on distributed optimization. A classical but powerful technique for distributed optimization without such knowledge is the *alternating direction method of multipliers*, surveyed in endnote [25] and applied to modern power system scenarios in endnote [26]. Many distributed optimization techniques are based on *consensus algorithms*, which essentially compute averages of multiple agents' parameters [27] and are being increasingly applied in power systems and a number of other fields, cf. [28, 29]. Unlike the above example, these are numerical algorithms that assume only local knowledge of network structure and are mechanistically applicable to broad classes of optimization problems.

7.3   More game theory

Section 6.1.3 introduced game theory and identified a number of its shortcomings, which included computational complexity and unrealistic modeling assumptions. Section 6.3 used game theory to analyze market power in power systems but had to trade many modeling details for analytical and computational tractability. We now discuss extensions of game theory that make it relevant to some modern scenarios, as well as some constructive tools for combatting market power.

7.3.1   Dynamic games

As discussed in Section 6.2.2, multi-period markets are becoming more and more relevant as intermittent renewable energy sources increase the need for regulation, and storage imposes dynamic energy constraints across consecutive time periods. History teaches us that this new economic dimension could become a new channel for market power if there is any possibility of further profit. To understand such vulnerabilities, we must apply game theory as in Section 6.3. The combination of game theory and dynamics places us in the realm of *dynamic games* [30]. Given the complexity issues of standard game theory discussed in Section 6.1.3, it is no surprise that adding dynamics doesn't make things any easier.

Consider the following energy storage game, which is a dynamic, stochastic version of the capacitated price competition from Section 6.3.3. At each time period, a random imbalance appears, w^t, which the system operator would like to absorb with storage at the cheapest cost. At each time, each energy storage submits a price bid, p_i^t, and w_t is allocated price-wise amongst the storages up to their positive or negative capacities at that time. In the two-player version, storage i has the dynamics

$$s_i^{t+1} = s_i^t + u_i^t,$$

where

$$u_i^t = \begin{cases} \min\left\{w^t, c_i - s_i^t\right\} & \text{if } p_i^t \le p_{-i}^t, \; w_t \ge 0 \\ \min\left\{\max\left\{w^t - c_{-i} + s_{-i}^t, 0\right\}, c_i - s_i^t\right\} & \text{if } p_i^t > p_{-i}^t, \; w_t \ge 0 \\ \max\left\{w^t, -s_i^t\right\} & \text{if } p_i^t \le p_{-i}^t, \; w_t < 0 \\ \max\left\{\min\left\{w^t + s_{-i}^t, 0\right\}, -s_i^t\right\} & \text{if } p_i^t > p_{-i}^t, \; w_t < 0 \end{cases}.$$

Player i then maximizes its net profit,

$$\underset{p_i}{\text{maximize}} \; \mathbb{E}\left[\sum_t p_i^t \left|u_i^t\right|\right],$$

where the absolute value is taken because all imbalances are rewarded.

Beyond being extremely difficult to analyze as is, this model begs a number of questions. Foremost, what is the right market format? How many stages ahead can each player realistically think; the most natural approach would thus be to start with the simplest, two-period case and proceed from there. Do both players have the same information about w_t, and does that information improve over time?

A potentially more fruitful approach is to restrict ourselves to linear quadratic games, in which a linear system dynamically couples the decisions of each player. Although limited in their modeling capability, linear quadratic games can admit far more tractable solutions, which are related to those for the linear quadratic regulator in Section 4.2.2.

7.3.2 Mechanism design

The game theoretic analyses of Section 6.3 give insight into how markets can be abused but offer little remedy besides case-by-case comparisons. *Mechanism design* seeks to provide constructive tools for steering markets (and other strategic situations) toward better outcomes [31–33]. In more technical terms, it is the problem of designing games to have better Nash (and other) equilibria. *Algorithmic game theory* is a related field concerned with the computational complexity of tasks like computing Nash equilibria and designing mechanisms for improving them [33]. Because we know that even basic problems like computing the equilibria of a two-player game are computationally diffi-cult [34], it is somewhat surprising that some very general and tractable approaches do exist. Perhaps the most well known is the *Vickrey-Clarke-Groves mechanism* [35–37], which we describe now.

Recalling our game theory notation from Section 6.1.3, let $\mathcal{S}$ be the set of all players' strategies, and let $u_i(s)$ be the monetary utility player i derives from $s \in \mathcal{S}$. Define $v_i(s)$

to be player i's declared utility; for example, a supply function reported to the system operator for use in economic dispatch. Our objective is to induce players to honestly report their utilities, so that $v_i = u_i$.

We define the *social choice function* to be the actions dictated by the players' declared utilities, $s = \mathcal{F}(v)$. For instance, in the case of economic dispatch, $\mathcal{F}$ maps the supply functions reported by the market participants to their power outputs. A central authority assesses a payment, $\mathcal{P}_i(v)$, on each player, for example a tax paid to the system operator prior to economic dispatch.

The Vickrey-Clarke-Groves mechanism is defined by the payments

$$\mathcal{P}_i(v) = \min_{s \in \mathcal{S}} \sum_{j \neq i} v_j(s) - \sum_{j \neq i} v_j(\mathcal{F}(v)). \tag{7.2}$$

The first term is the best possible system outcome or *social welfare* when player i is nonexistent, and the second term is the actual welfare of all players but i. $\mathcal{P}_i(v)$ therefore is the aggregate financial improvement seen by all players when i is removed.

We say that the Vickrey-Clarke-Groves mechanism is *incentive compatible* because the payments $\mathcal{P}_i(v)$ make honesty optimal for each player. More precisely, here honesty is a *dominant strategy equilibrium*, which means it is optimal regardless of the others' decisions. Assuming each agent is minimizing, this is stated mathematically as

$$u_i(\mathcal{F}(u_i, v_{-i})) - \mathcal{P}_i(u_i, v_{-i}) \leq u_i(\mathcal{F}(v)) - \mathcal{P}_i(v) \quad \text{for all} \quad v.$$

Note that incentive compatibility does not rule out the possibility of other equilibria, which may not have the same desirable properties. Also observe that every dominant strategy equilibrium is a Nash equilibrium as defined in Section 6.1.3, but a Nash equilibrium is generally not a dominant strategy equilibrium.

Actually, only the latter term in (7.2) is necessary for incentive compatibility. The first term, which is known as the Clarke pivot rule, helps shrink the payments relative to the utilities themselves and under certain conditions imparts additional desirable properties. Finally, we remark that incentive compatibility generally leaves the central authority budget imbalanced [38], which may be unacceptable in some contexts. Nevertheless, the Vickrey-Clarke-Groves mechanism and other tools from mechanism design are promising approaches to market power in the scenarios discussed in Chapter 6 and have already seen application to some problems in power systems like economic dispatch [39] and regulation [40].

References

[1] R. Billinton and R. N. Allan, *Reliability Evaluation of Power Systems*, 2nd ed. Springer, 1996.

[2] M. R. Patel, *Wind and Solar Power Systems: Design, Analysis, and Operation*. CRC Press, 2006.

[3] P. Kall and S. W. Wallace, *Stochastic Programming*. Chichester: John Wiley & Sons, 1994.

[4] J. R. Birge and F. V. Louveaux, *Introduction to Stochastic Programming*. Springer, 1997.

[5] F. Bouffard and F. Galiana, "Stochastic security for operations planning with significant wind power generation," *IEEE Transactions on Power Systems*, vol. 23, no. 2, pp. 306–316, 2008.

[6] A. Papavasiliou, S. Oren, and R. O'Neill, "Reserve requirements for wind power integration: A scenario-based stochastic programming framework," *IEEE Transactions on Power Systems*, vol. 26, no. 4, pp. 2197–2206, 2011.

[7] S. Boyd and L. Vandenberghe, *Convex Optimization*. New York: Cambridge University Press, 2004.

[8] R. Jabr, "Robust self-scheduling under price uncertainty using conditional value-at-risk," *IEEE Transactions on Power Systems*, vol. 20, no. 4, pp. 1852–1858, 2005.

[9] A. Ben-Tal and A. Nemirovski, "Robust convex optimization," *Mathematics of Operations Research*, vol. 23, no. 4, pp. 769–805, 1998.

[10] D. Bertsimas and M. Sim, "The price of robustness," *Operations Research*, vol. 52, no. 1, pp. 35–53, 2004.

[11] A. Ben-Tal, L. Ghaoui, and A. Nemirovski, *Robust Optimization*, ser. Princeton Series in Applied Mathematics. Princeton University Press, 2009.

[12] K. Zhou, J. C. Doyle, and K. Glover, *Robust and Optimal Control*. Prentice Hall, 1996.

[13] S. Boyd, L. El Ghaoui, E. Feron, and V. Balakrishnan, *Linear Matrix Inequalities in System and Control Theory*, ser. Studies in Applied Mathematics. Philadelphia: SIAM, 1994, vol. 15.

[14] G. Boukarim, S. Wang, J. Chow, G. Taranto, and N. Martins, "A comparison of classical, robust, and decentralized control designs for multiple power system stabilizers," *IEEE Transactions on Power Systems*, vol. 15, no. 4, pp. 1287–1292, 2000.

[15] Ibraheem, P. Kumar, and D. Kothari, "Recent philosophies of automatic generation control strategies in power systems," *IEEE Transactions on Power Systems*, vol. 20, no. 1, pp. 346–357, Feb. 2005.

[16] A. Conejo, J. Morales, and L. Baringo, "Real-time demand response model," *IEEE Transactions on Smart Grid*, vol. 1, no. 3, pp. 236–242, 2010.

[17] D. Bertsimas, E. Litvinov, X. Sun, J. Zhao, and T. Zheng, "Adaptive robust optimization for the security constrained unit commitment problem," *IEEE Transactions on Power Systems*, vol. 28, no. 1, pp. 52–63, 2013.

[18] J. Warrington, P. Goulart, S. Mariethoz, and M. Morari, "Policy-based reserves for power systems," *IEEE Transactions on Power Systems*, vol. 28, no. 4, pp. 4427–4437, 2013.

[19] N. Sandell, P. Varaiya, M. Athans, and M. Safonov, "Survey of decentralized control methods for large scale systems," *IEEE Transactions on Automatic Control*, vol. 23, no. 2, pp. 108–128, 1978.

[20] M. Rotkowitz and S. Lall, "A characterization of convex problems in decentralized control," *IEEE Transactions on Automatic Control*, vol. 51, no. 2, pp. 274–286, 2006.

[21] A. Venkat, I. Hiskens, J. Rawlings, and S. Wright, "Distributed MPC strategies with application to power system automatic generation control," *IEEE Transactions on Control Systems Technology*, vol. 16, no. 6, pp. 1192–1206, Nov. 2008.

[22] P. Shah and P. Parrilo, "$\mathcal{H}_2$-optimal decentralized control over posets: A state-space solution for state-feedback," *IEEE Transactions on Automatic Control*, vol. 58, no. 12, pp. 3084–3096, Dec 2013.

[23] A. Nayyar, A. Mahajan, and D. Teneketzis, "Decentralized stochastic control with partial history sharing: A common information approach," *IEEE Transactions on Automatic Control*, vol. 58, no. 7, pp. 1644–1658, July 2013.

[24] D. Bertsekas and J. Tsitsiklis, *Parallel and Distributed Computation*. Prentice-Hall, 1989.

[25] S. Boyd, N. Parikh, E. Chu, B. Peleato, and J. Eckstein, *Distributed Optimization and Statistical Learning Via the Alternating Direction Method of Multipliers*. Now Publishers, 2011.

[26] M. Kraning, E. Chu, J. Lavaei, and S. Boyd, "Dynamic network energy management via proximal message passing," *Optimization*, vol. 1, no. 2, pp. 70–122, 2013.

[27] J. Tsitsiklis, D. Bertsekas, and M. Athans, "Distributed asynchronous deterministic and stochastic gradient optimization algorithms," *IEEE Transactions on Automatic Control*, vol. 31, no. 9, pp. 803–812, Sept. 1986.

[28] A. Olshevsky and J. Tsitsiklis, "Convergence speed in distributed consensus and averaging," *SIAM Journal on Control and Optimization*, vol. 48, no. 1, pp. 33–55, 2009.

[29] A. Dominguez-Garcia, C. Hadjicostis, and N. Vaidya, "Resilient networked control of distributed energy resources," *IEEE Journal on Selected Areas in Communications*, vol. 30, no. 6, pp. 1137–1148, July 2012.

[30] T. Başar and G. Olsder, *Dynamic Noncooperative Game Theory*, 2nd ed., ser. Classics in Applied Mathematics, 23. Society for Industrial & Applied Mathematics, 1999.

[31] D. Fudenberg and J. Tirole, *Game Theory*. MIT Press, 1991.

[32] A. Mas-Colell, M. Whinston, and J. Green, *Microeconomic Theory*. Oxford University Press, 1995.

[33] N. Nisan, T. Roughgarden, E. Tardos, and V. V. Vazirani, *Algorithmic Game Theory*. New York: Cambridge University Press, 2007.

[34] C. Daskalakis, P. Goldberg, and C. Papadimitriou, "The complexity of computing a Nash equilibrium," *SIAM Journal on Computing*, vol. 39, no. 1, pp. 195–259, 2009.

[35] W. Vickrey, "Counterspeculation, auctions, and competitive sealed tenders," *The Journal of Finance*, vol. 16, no. 1, pp. 8–37, 1961.

[36] E. H. Clarke, "Multipart pricing of public goods," *Public Choice*, vol. 11, no. 1, pp. 17–33, 1971.

[37] T. Groves, "Incentives in teams," *Econometrica: Journal of the Econometric Society*, vol. 41, no. 4, pp. 617–631, 1973.

[38] J. Green and J. Laffont, *Incentives in Public Decision Making*. North-Holland, Amsterdam, 1979.

[39] R. Johari and J. N. Tsitsiklis, "Parameterized supply function bidding: Equilibrium and efficiency," *Operations Research*, vol. 59, no. 5, pp. 1079–1089, 2011.

[40] J. A. Taylor, A. Nayyar, D. S. Callaway, and K. Poolla, "Consolidated dynamic pricing of power system regulation," *IEEE Transactions on Power Systems*, vol. 28, no. 4, pp. 4692–4700, 2013.

Index